Basic Radio:
Principles and Technology

Ian Poole BSc(Eng) CEng MIEE

 Newnes

Newnes
An imprint of Butterworth-Heinemann
Linacre House, Jordan Hill, Oxford OX2 8DP
225 Wildwood Avenue, Woburn, MA 01801-2041
A division of Reed Educational and Professional Publishing Ltd

℞ A member of the Reed Elsevier plc group

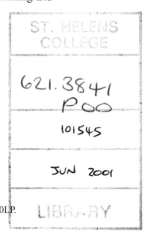
First published 1998
Reprinted 1999, 2000

© Ian Poole, 1998

British Library Cataloguing in Publication Data
A catalogue record for this book is available from the British Library.

Library of Congress Cataloguing in Publication Data
A catalogue record for this book is available from the Library of Congress.

ISBN 0 7506 2632 1

Composition by Genesis Typesetting, Rochester, Kent
Printed and bound in Great Britain by Biddles Ltd
www.biddles.co.uk

British Trust for
Conservation Volunteers

FOR EVERY TITLE THAT WE PUBLISH, BUTTERWORTH-HEINEMANN
WILL PAY FOR BTCV TO PLANT AND CARE FOR A TREE.

Contents

1 Radio today, yesterday and tomorrow

The benefits of radio technology are an integral part of our everyday life. Over the past decades we have come to accept radio, taking it for granted and not realizing how much it contributes to the running of modern society.

Radio performs many functions today, but the most obvious must be domestic broadcasting. Most homes have a variety of radio sets, ranging from the relatively simple and cheap portable radios, through the more sophisticated car radios to the high fidelity systems. All of these sets offer a high degree of performance which is the result of use and development over many years. Even so, the performance of radio sets is being improved and new facilities are being introduced all the time. The introduction of wide-band FM represented a major improvement when it was first introduced in the early 1950s. Stereo and RDS (Radio Data System) are two more which are in use today, and digital audio broadcasting has started. This brings true CD quality to radio broadcasts as well as allowing data to be broadcast at the same time.

Cellular phones are another benefit which have been brought about by the development of radio. They have enabled people to make phone calls or to be contacted even when they are away from home or the office. Their flexibility has meant that since their introduction in the UK in 1984 there has been a phenomenal growth in this market. Phones are available very cheaply, and even the line rental and call charges are falling as the number of operators increases. This means that a wide variety of people are able to use these phones. Initially they were mainly for business use. Now with the falling charges the number of private users is increasing.

The introduction of the cellular phone systems has done much to advance radio technology. Since their first introduction the sizes of phones have fallen. The first portable sets required the use of large batteries, whereas now

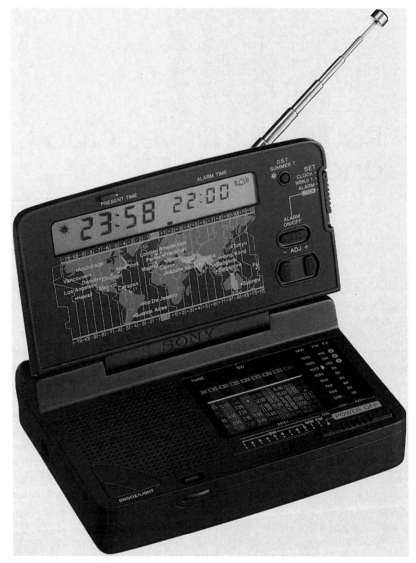

Figure 1.1 *A modern 'World Band' radio (Courtesy of Sony UK Ltd)*

the phones are truly hand-held and can be slipped into any pocket. Achieving this has required a large amount of development. More areas of the phones have been integrated onto silicon chips. Now low power consumption, ultra high frequency, radio frequency integrated circuits are commonplace where a few years ago they were few and far between.

Figure 1.2 *A pocket sized cellular phone (Courtesy of Sony UK Ltd)*

Wireless communications have also enabled information to be sent to and from all parts of the world more quickly. Now it is possible to send and receive data, faxes, or voice from almost anywhere in the world very easily. A good example of the way information can be sent more swiftly is shown in the way news is received from around the world almost instantaneously. This can be seen by comparing the news reports of the Second World War, and the Gulf War of 1991. In the 1940s it took several hours for recorded reports from the front-line correspondents to reach the broadcasting stations. Film reports took several days or weeks before they were seen as newsreels in the local cinemas. Now with satellite communications, modern video techniques and television, things happen much faster. In the Gulf War the action was seen by millions in their homes as it happened. Fortunately there are many more less hostile examples of how radio has helped improve the standards of modern life. Here radio has enabled many people to be rescued by the emergency services far more efficiently than if there were no radio present.

There has been a vast amount of investment needed to achieve the high standards of communication needed today. Satellites, microwave links, cellular phone base stations, sophisticated radio data links are but a few which are needed to support today's requirements. To achieve this a number of different types of technology are needed: antennas, transmitters, and receivers as well as several other items are all needed to make up the complete system. This provides a great amount of variety for anyone looking to take up radio as a career or hobby.

Before progressing into today's radio technology it is interesting to put it in perspective and look at the way in which radio developed. Over more than a hundred years, many people have helped it reach its current state of development. The story of its development is fascinating and can be traced back to people like Volta, Ampere and Ohm who made some of the first discoveries associated with electricity; but the story of radio itself begins with a brilliant Scot named Maxwell.

The story of radio

Maxwell was born in 1831 and entered Edinburgh University when he was only sixteen. After graduating, he spent time at a number of universities, but it was when he was at King's College in London that he undertook most of the work into electromagnetic theory for which he is famous. He published three main papers between 1855 and 1864, and finally summarized his work in a book entitled *Treatise on Electricity and Magnetism*. His work proved the existence of an electromagnetic wave. However, much of Maxwell's work was theoretical and he was never able to demonstrate the presence of electromagnetic waves in practice. Sadly, Maxwell died at the early age of 48, and the work he started was left to others to continue.

The quest for the electromagnetic wave took many years. A number of people, including Edison and Henry, came close to discovering it. However, it was a German named Heinrich Hertz to whom the honour fell.

Hertz performed a wide variety of experiments to prove the existence of these new waves. He also gave a number of demonstrations and lectures. In one of these he used an induction coil connected to a loop of wire in which two large spheres were placed slightly apart. The induction coil generated a large voltage in the circuit causing a spark to jump across the gap. In turn this caused a spark to jump across the gap of a similar coil with two spheres placed within a few metres of the transmitter.

Using other equipment Hertz managed to prove many of the basic properties of these waves. He showed that they had the same velocity as light, and they were refracted and reflected in the same way. As Hertz had discovered the waves, they soon became known as Hertzian waves.

With the existence of the electromagnetic wave firmly established it did not take long before people started to think of using them for communicating. However, to be able to achieve this it was necessary to be able to have a much better way of detecting them. This came in the form of the coherer. It was initially designed by a Frenchman named Edouard Branly in 1890. He discovered that the resistance of a glass tube filled with metal filings fell from a very high resistance to a few hundred ohms when the filings cohered or clung together when an electrical discharge took place nearby. It could then be reset quite easily by a small tap on the glass.

After its initial discovery the coherer was greatly improved and popularized by an English scientist named Sir Oliver Preece. Such were his improvements that in 1894 he was able to detect signals from a transmitter a few hundred yards away.

It was around this time that a young Italian named Marconi started to experiment with Hertzian waves. His drive, intuition and business sense enabled this new science to progress much faster than it would otherwise have done.

Marconi was born in Bologna in Northern Italy in 1874, receiving a private education during his early years. Despite his parents' expectations he failed to gain a place at Bologna University. Fortunately he had a growing interest in science and he was encouraged by a family friend who was a lecturer at the university. Marconi was allowed to sit through his lectures and consequently he discovered about the new Hertzian waves.

Marconi quickly became interested and soon started to perform many experiments in the attic of his parents' house. Early in the summer of 1895 he managed to transmit a signal a distance of a few yards. By the end of the summer he had succeeded in receiving a signal over 2 kilometres away from the transmitter.

Even at this early stage Marconi was able to see the commercial possibilities. Accordingly he approached the Italian Ministry of Posts with his ideas for wireless communications, but his proposal was turned down. It was this rejection which caused Marconi to come to England in 1896.

On his arrival he soon set to work and filed a patent for a system of Wireless Telegraphy using Hertzian waves which was granted on 2 June 1896. In England there was considerably more interest for Marconi's work. Soon he was introduced to a man named William Preece who was the chief engineer of the Post Office.

Marconi gave some preliminary demonstrations of his equipment in the laboratory and then he set up a transmitter and receiver on the roofs of some buildings in London a few hundred yards apart. The success of this demonstration promoted a further demonstration on Salisbury Plain in September 1896 when representatives of the Post Office together with others from the War Office and the Navy were present.

The Navy saw the possibilities of using wireless equipment for communication at sea and they showed considerable interest. However, Marconi also started to sell his equipment to other maritime users. Initially the take-up was slow, but soon other organizations such as Lloyd's endorsed its use as a method of sending distress signals and very quickly more vessels were fitted with Marconi's equipment.

Not satisfied with supplying equipment for maritime use, Marconi also started to investigate its use for providing a long distance communications link. Initial experiments sending a message across the English Channel proved the possibilities and gave valuable propaganda. However, the main goal was to be able to send a message across the Atlantic. This was not an easy task. Many difficulties needed to be overcome, but with the help of his team, consisting of Vyvyan, Professor Fleming of University College London, Paget, and Kemp, stations at Poldhu in Cornwall and Cape Cod were soon established. Unfortunately the antennas were destroyed in storms and it was decided that their design should be changed. At the same time the site at Cape Cod was abandoned in favour of one at St Johns in Newfoundland. Finally on 12 December 1901 the first transmission was received when the letter 'S' was detected in the receiver and Marconi became a legend in his own time.

This was a major success, and it ably proved the value of wireless as a means of providing long distance communications, but it also highlighted the many shortcomings of the systems in use at the time. The most noticeable problem was the lack of sensitivity of the detector in the receiver.

It was this problem which occupied much of the thought of Professor Ambrose Fleming. He had played a significant part in the transatlantic success, designing many parts of the equipment including the transmitter. For a number of years he had known about a phenomenon called the Edison Effect. It was found that when a second element was introduced into a light bulb, current would only pass in one direction between the two elements. Both Edison and Fleming had observed the effect, but neither had been able to think of a use for it. However, one day as Fleming was walking down Gower Street in London, he had what he later described as 'a sudden very happy thought'. He realized the possibilities of using the Edison Effect in detecting Hertzian waves. Fleming quickly instructed his assistant to set up an experiment to see if the idea worked and to their delight it proved very successful.

Fleming patented the idea on 16 November 1904, calling the device his oscillation valve because of its unidirectional properties with electrical oscillations. In many respects this was the first major component of the electronic era.

The oscillation valve consisted of a filament heated to white heat. This caused electrons to leave the filament because of an effect known as

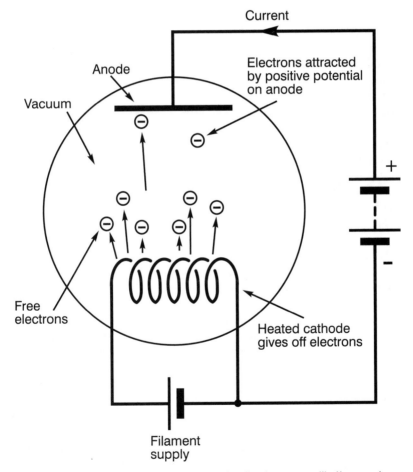

Figure 1.3 *The idea behind Fleming's diode or oscillation valve*

thermionic emission. This makes the filament positively charged and the electrons are attracted back. However, if another element called an anode is introduced and is given a positive charge then it will attract the electrons away from the heated filament or cathode.

Fleming used this effect to rectify the radio or wireless signals so that the modulation of the alternating current signals was detected. The signals could then be displayed on a meter or heard in headphones.

The next major development was made by an American named Lee de Forest. He saw the possibilities of Fleming's diode and started to work on copies of the device. In an attempt to be able to circumvent Fleming's patents he investigated a number of other ideas, placing a third electrode into the envelope between the cathode and anode. He called this electrode a grid because it consisted of a wire mesh or grid through which the electronics could pass.

de Forest took out a number of patents for his idea, but it was not until around 1907 that he developed the new valve, which he called an Audion, sufficiently for it to be used properly. Unknown to him at this time, de Forest had struck on an idea of immense importance, but initially the triode valve was only used as a detector. It took a number of years before it was used as an amplifier.

After their introduction, valves were not widely used. They were expensive to buy and were also very expensive to run. At this time valve technology was such that each valve required a separate battery to supply its filament, and as they consumed an amp or more each, the batteries did not last very long. This resulted in other developments coming to the fore. The main one of these was the crystal detector.

These detectors consisted of a crystal which was usually made out of a substance called galena. A thin wire was positioned onto it to make a point contact which acted as a rectifier to detect the signals. These detectors became commonly known as 'cat's whiskers', and were widely used because they were very cheap. Unfortunately they were not very reliable, and often the position of the whisker had to moved to a better position on the crystal.

It was only after the crystal detector was well established that the Audion was used for amplifying signals. It took until 1911 for it to be used in this role, and de Forest quickly exploited the idea when he started to build amplifiers for use as telephone repeaters.

As the triode valve started to be used as an amplifier its property to act as an oscillator was very quickly discovered. In fact these early valves were often difficult to stop oscillating, especially at radio frequencies. This was put into good use in a number of ways. As far as radio reception was concerned it soon led to the development of the regenerative receiver. Here positive feedback in the circuit was used to increase the sensitivity and selectivity, although the circuit was maintained just before the point of oscillation for the

Figure 1.4 *A cat's whisker detector assembled on a 'bijou' crystal receiver of 1923 (Courtesy of G. C. Arnold Partners)*

best performance. The American Edwin Armstrong is normally credited with this invention, although others, including de Forest, Irving Langmuir in the USA, and Alexander Meissner in Germany, also discovered the idea at about the same time.

In 1914, war broke out. With this, wireless research on both sides of the conflict took on a new degree of urgency as its importance was recognized. In developing new and better receivers many famous engineers, including H. J. Round, Latour, and Edwin Armstrong on the Allied side and W. Schottky for the Germans, devoted vast amounts of research effort into the development of better techniques. Although the idea of the regenerative receiver had given significant improvements over anything which was previously available, receivers still lacked the performance which was sought. Many of the limitations were caused by the poor performance of the valves. High inter-electrode capacitances meant that at frequencies above a few hundred kilohertz the valves lacked any usable gain and became very unstable. Even at low frequencies they were very noisy.

H. J. Round made some significant developments in valve technology to try to overcome the problem. However, it was a Frenchman named Lucien Levy who opened the gate to the real solution. In 1917 he devised a system of reception which he claimed would completely eliminate 'parasites and ordinary interference'. To achieve this he used the principle of beats to convert signals from a high frequency down to a lower one where the variable filters could be made more selective.

The American Edwin Armstrong later used the same basic principle in his receiver. He used a variable frequency local oscillator to convert the incoming signals down in frequency. A fixed frequency filter was then used to give the selectivity. By varying the frequency of the oscillator the receiver could be tuned from one frequency to the next. This idea had a number of advantages. Using a low frequency for the filter gave better selectivity, and by keeping its frequency fixed, several stages could be placed in series to give even better performance. He also included a significant amount of amplification at this lower intermediate frequency stage, where better gain could be achieved from the valves.

Armstrong patented his idea for the receiver he called a supersonic heterodyne, or superhet receiver, on 30 December 1918. The superhet was too late to see any real use in the war or for a number of years thereafter. The main reason for this was that it used more valves than its regenerative counterparts, and for normal broadcast reception the additional performance was not needed.

At this time most long distance communications were undertaken on very long wavelengths. As a result amateur experiments were relegated to the short wave bands with wavelengths shorter than 200 metres. These bands were thought to be of little use, but undeterred by this radio amateurs sought to disprove this and make long distance contacts.

In America there were a large number of radio amateurs who were permitted to use relatively high powers, and reports of long distance contacts soon started to appear. With this in mind, the possibility of short wave communication across the Atlantic was thought to be a possibility. The first attempts were made in December 1921, with British stations listening for their American counterparts who could use much higher powers. These first attempts proved to be unsuccessful; however, the following winter many stations were heard. Then in November 1923, the first two way contact was made between a French station and an American in Connecticut. With this barrier broken regular transatlantic communications started to be made. Contacts were also made over further distances; one between London and Dunedin in New Zealand was made in October 1924.

While radio amateurs took the limelight, commercial interests also began to investigate the possibilities of the short wave bands. A telephone circuit was set up between Hendon in North London and Birmingham in 1921, and later that year another was inaugurated between Southwold in Sussex and Zandvoort in Holland.

Seeing the success of all these projects Marconi decided to make further investigations into the properties of these bands. He set up a 12 kW transmitting station at Polhu, the site of his long wave transmitter in 1901, and he used receivers on his yacht *Elettra* to monitor the transmissions. He found that signals up to St Vincent in the West Indies (a distance of 2300 miles) could be received at sufficient strength to support a telephone channel.

Many other tests and experiments were performed, and slowly the short wave bands started to be used for international traffic. By the late 1920s radio was providing a valuable method of providing international telephone circuits. Links were even opened up to ocean liners to provide instant communication for passengers who were travelling on board.

Wireless was not only used for the transmission of commercial traffic. Some early broadcasts of entertainment were made, but the first regular broadcasts were made by the Marconi station in Chelmsford. This station opened on 23 February 1920, making a daily broadcast of half an hour of speech, music and news. About four months later on 20 June history was made when the Australian singer, Dame Nellie Melba, made the first broadcast by a professional artist. The transmission made the front pages of the newspapers, but interference caused by the transmissions forced the closure of the station.

Public opinion soon meant that the decision had to be reversed. On 14 February 1922 the Marconi station licensed as 2MT was allowed to broadcast a 15 minute programme. This proved to be very successful and on 11 May a station commenced broadcasting from Marconi's headquarters in the centre of London. This was the famous station 2LO which was taken over by the

Figure 1.5 *A Radio Instruments of London V.4.A receiver of 1923 (Courtesy of G. C. Arnold Partners)*

British Broadcasting Company (later to become the British Broadcasting Corporation or BBC) in November of that year.

With broadcasting now established more stations were set up. Before long most areas of the British Isles had their own station. Broadcasting increased in other countries as well. In the USA the rise in the number of stations was far greater, placing far greater requirements on receiving equipment, and encouraging the use of Armstrong's superhet receiver.

The short wave bands also started to be used for international broadcasting. Early experiments were undertaken by an American station owned by the Westinghouse Company. Transmitting from Pittsburg on a wavelength of 62.5 metres, the station (KDKA) was successfully received in Britain and actually rebroadcast. In 1927 an experimental short wave broadcasting service was set up in Britain. The first station was located in Chelmsford and this service continued until 1932 when the BBC Empire Service transmitting from a site in Daventry was opened. Its aim was to keep all parts of the British Empire with London by being able to receive at least one transmission a day.

With short wave broadcasting well-established attention was soon turned to the possibility of using ultra short waves for communication. Initially this posed problems with suitable devices for operation at these frequencies. However, these problems were slowly overcome and in March 1931 Standard Telephones and Cables Limited gave a demonstration of telephony across the Straits of Dover using a wavelength of just 18 centimetres. The British Post Office also installed some services on these frequencies. In 1932 they opened a telephone circuit across the Severn on 5 metres which saved having to use a much longer cable route through the Severn Tunnel.

The idea of modulating the frequency of a signal instead of its amplitude had been known from the earliest days of radio. Early experiments had been undertaken in 1902 to investigate its viability. Later ones also treated it in a similar way to AM. In attempts to reduce the levels of background noise and interference the bandwidth had been reduced and this had only served to impair its performance.

However, Edwin Armstrong who was already a millionaire from his previous inventions, turned traditional thinking around. Instead of reducing the bandwidth, he increased it. Early interest in the idea was very limited and so Armstrong set up his own transmitter and receiver to demonstrate the idea in 1935. Using a frequency of 110 MHz he was able to conclusively prove that a wide-band FM system gave significant improvements over a traditional AM one. Also using higher frequencies he would be able to reduce the congestion on the long and medium wave bands which were already overcrowded.

Despite the success of the demonstration, interest from broadcasters was very limited. Undeterred by this, Armstrong set up his own station which opened in July 1939. Soon other stations saw the possibilities of improved reception and within six months over 140 applications had been submitted in the USA for new VHF FM stations.

The Second World War delayed the spread of wide-band FM, and it was not until 1954 that the BBC launched its first FM service. Since then the growth of wide-band FM has been enormous. It now attracts greater audiences than the traditional AM broadcasts because of its improved sound reproduction and resilience to noise and fading.

Communications will always be a very important use for radio. During 1946, the scientist and science fiction writer Arthur C. Clarke published a revolutionary article in the magazine *Wireless World*. In this he proposed the use of three orbiting satellites to give worldwide communication. While his ideas were remarkably sound, the technology did not exist to put them into practice. However, during the Second World War the Germans had made major advances into rocket technology. These were the first steps which enabled rockets to put satellites into space, and during the 1950s many further advances were made.

Many of the first attempts at putting satellites into space failed dramatically, but on 4 October 1957 the Russians succeeded in placing a football sized satellite into a low orbit. Named *Sputnik 1* it radiated a distinctive bleep which was heard by many professionals and amateur listeners all over the world.

Sputnik was the first major milestone in the satellite age. Soon others followed. In 1960 the Americans launched *Echo*, a 100 foot diameter aluminium coated balloon. This could easily be seen from earth as it passed overhead and it enabled radio signals to be reflected to provide extended distances for radio communications. This was only the first step in making a proper communications satellite. In 1962 *Telstar* was launched. This was a private venture funded by AT&T and on 23 July 1962 it succeeded in providing the first live transatlantic pictures relayed by satellite.

After the success of *Telstar* developments progressed at a rapid rate and many more satellites were launched. Soon they were providing an enormous variety of services from communications, to weather monitoring, and direct broadcasting to navigation.

Satellites were by no means the only improvement for communications. With electronic technology progressing at an ever increasing rate the need for personal communications was seen as the next major goal. Accordingly in 1984 two cellular mobile phone systems were launched in Britain. One system named Cellnet was owned jointly by British Telecom and Securicor, the other was Vodafone, then owned by Racal. These systems soon became an enormous success as the need for mobile communications was realized by the business community. After the launch of analogue systems newer digital systems came on line providing improved performance. Additional developments in miniaturization of electronics have enabled highly sophisticated phones to be made smaller than the normal telephone handsets.

Since the first radio or wireless transmissions were made the field of radio has grown by enormous proportions. Now, more than ever, it is a cornerstone to our everyday life, both at home and at work. With technology progressing as it is, radio is likely to become even more important in the years to come, giving more flexibility and higher quality services. It is also likely to find an even greater variety of uses, many of which have not even been conceived yet.

Radio tomorrow

There is a very large amount of development being undertaken in a wide variety of areas associated with radio. In recent years the flexibility which radio systems offer over cable based ones has started to be exploited. Cellular telephones have seen a phenomenal growth rate in recent years. In view of the fact that costs are falling, their use is now open to a wide variety of people, whereas initially they were mainly used for businesses.

In the coming years it is likely that system coverages will be increased. More countries will start to use them, and many more facilities will become available. The global system (GSM) is also likely to see an increase in use, especially among business users who need to travel. Also, digital systems in general are becoming more widespread to overcome the problem of eavesdropping by people with scanners.

Broadcasting is also being developed to ensure that it meets the needs of the twenty-first century. New methods of broadcasting digital sound are being introduced. These are capable of providing CD quality sound, even to listeners in cars where previously the FM services were lacking. In addition to an improvement in the sound quality, the digital nature of the system opens up the possibility of introducing many new facilities to meet the needs of the listeners of the future.

Apart from introducing new services, many other radio developments are focused on pushing forwards the boundaries of technology. The pressure on the amount of radio spectrum which is available means that higher frequencies need to be used. To meet this need new semiconductor devices are being developed which are capable of operating at much higher frequencies than before. Although these devices may not be widely available for a number of years, devices which were at the forefront of technology some years ago are now in widespread use enabling many of the systems which we take for granted to function.

These developments represent just a few of the areas where radio technology is moving forwards. However, they are not the only areas. In view of the importance of radio in today's society, there are many areas where new developments are being sought to enable new radio applications to be implemented and others to enable it to operate more effectively. All of them are an indication of the importance which the electronics industry attaches to radio. This means that it is a technology for the future.

2 Radio waves and propagation

The properties of radio waves and the way in which they travel or propagate are of prime importance in the study of radio technology. These waves can travel over vast distances enabling communication to be made where no other means is possible. Their use enables communication to be established over distances ranging from a few metres to many thousands of miles, permitting telephone conversations and many other forms of communication to be made with people on the other side of the world using short wave propagation or satellites. Radio waves can be received over even greater distances. Radio telescopes pick up minute signals from sources many light years away.

Another property of radio waves is that they spread out in all directions from the transmitting antenna. This enables them to be used for broadcasting. Nowadays most homes have several radio sets and most countries have a variety of stations which can be picked up. In Europe and North America there is a very large choice of stations which can be received, and this proves the value and success of the medium.

However, to understand more about radio it is necessary to have a comprehension of radio waves and how they propagate. This can be a very interesting topic, as their propagation varies with frequency. Sometimes they may travel around the earth, whereas at other frequencies they can travel into outer space enabling them to reach satellites or space craft.

Radio waves

Radio signals are a form of electromagnetic wave. They consist of the same basic type of radiation as light, ultraviolet and infrared rays, differing only in their wavelength and frequency. These waves are quite complicated in their make-up, having both electric and magnetic components which are

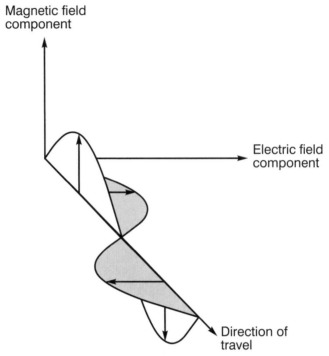

Magnetic field
component

Electric field
component

Direction of
travel

Figure 2.1 *An electromagnetic wave*

inseparable. The planes of these fields are at right angles to one another and
to the direction of motion of the wave. These waves can be visualized as
shown in Figure 2.1.

The electric field results from the voltage changes occurring in the antenna
which is radiating the signal, and the magnetic changes result from the current
flow. It is also found that the lines of force in the electric field run along the
same axis as the antenna, but spreading out as they move away from it. This
electric field is measured in terms of the change of potential over a given
distance, e.g. volts per metre, and this is known as the field strength.

There are a number of properties of a wave. The first is its wavelength.
This is the distance between a point on one wave to the identical point on the
next, as shown in Figure 2.2. One of the most obvious points to choose is the
peak as this can be easily identified although any point is acceptable.

The second property of the electromagnetic wave is its frequency. This is
the number of times a particular point on the wave moves up and down in a
given time (normally a second). The unit of frequency is the hertz and it is
equal to one cycle per second. This unit is named after the German scientist
who discovered radio waves. The frequencies used in radio are usually very
high. Accordingly the prefixes kilo, mega, and giga are often seen: 1 kHz is

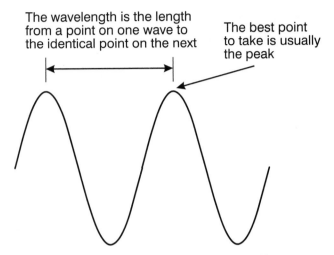

The wavelength is the length from a point on one wave to the identical point on the next

The best point to take is usually the peak

Figure 2.2 *The wavelength of an electromagnetic wave*

1000 Hz, 1 MHz is a million hertz, and 1 GHz is a thousand million hertz, i.e. 1000 MHz. Originally the unit of frequency was not given a name and cycles per second (c/s) were used. Some older books may show these units together with their prefixes: kc/s; Mc/s etc. for higher frequencies.

The third major property of the wave is its velocity. Radio waves travel at the same speed as light. For most practical purposes the speed is taken to be 300 000 000 metres per second although a more exact value is 299 792 500 metres per second.

Frequency to wavelength conversion

Many years ago the position of stations on the radio dial were given in terms of wavelengths. A station might have had a wavelength of 1500 metres. Today stations give out their frequency because this is far easier to measure. A frequency counter can be used to measure this very accurately, and with today's technology their cost is not unreasonable. It is very easy to relate the frequency and wavelength as they are linked by the speed of light as shown:

$$\lambda = \frac{c}{f}$$

where λ = the wavelength in metres
f = frequency in hertz
c = speed of radio waves (light) taken as 300 000 000 metres per second for all practical purposes

Taking the previous example the wavelength of 1500 metres corresponds to a frequency of 300 000 000/1500 or 200 thousand hertz (200 kHz).

Polarization

Electromagnetic waves can be polarized. The polarization of any wave is the plane in which it is vibrating. Electromagnetic waves consist of electric and magnetic components in different planes, and to determine the polarization of the wave the plane of the electric component is taken.

One of the most obvious examples of polarization is found with light waves. Polaroid lenses are able to allow light of a particular polarization through. This can be very useful on sunny days when reflections from surfaces can cause glare. As the reflected light has a particular polarization it is possible to reduce the glare by using polaroid lenses to cut out the reflections.

The polarization of a radio wave can be very important because it is found that antennas are sensitive to polarization, and generally only pick up or transmit a signal with a particular polarization. It is also found that the path travelled by a radio signal can affect its polarization, for example if it is reflected.

Radio spectrum

Radio waves are one form of electromagnetic radiation. They have the lowest frequency, and hence the longest wavelengths. Above the radio spectrum, other forms of radiation can be found. These include infrared radiation, light, ultraviolet and a number of other forms of radiation as shown in Figure 2.3.

The radio spectrum itself covers a vast range. At the bottom end of the spectrum there are signals of just a few kilohertz, whereas at the top end new

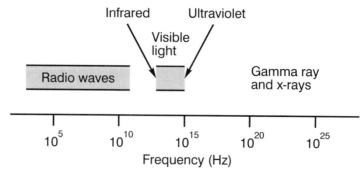

Figure 2.3 *Electromagnetic wave spectrum*

semiconductor devices are being developed which operate at frequencies of a hundred gigahertz and more. Between these extremes lie all the signals with which we are familiar. It can be seen that there is a vast amount of spectrum space available for transmissions. To make it easy to refer to different portions of the spectrum, designations are given to them as shown in Figure 2.4. It can be seen from this that transmissions in the long wave broadcast band (140.5 to 283.5 kHz) available in some parts of the world fall into the low frequency or LF portion of the spectrum. There are also a number of other types of transmission which are made here. For example, there are a number of navigational beacons which transmit on frequencies around 100 kHz or less.

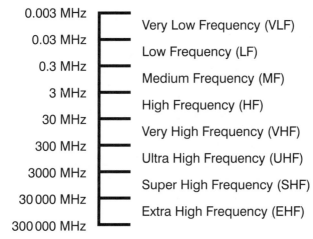

Figure 2.4 *The radiofrequency spectrum*

Moving up in frequency, the medium wave broadcast band falls into the medium frequency or MF portion of the spectrum. Above this broadcast band is often where the lowest frequency short wave bands start. Here there is an amateur radio band together with allocations for maritime communications.

Between 3 and 30 MHz is the high frequency or HF portion. Within this frequency range lie the real short wave bands. Signals from all over the world can be heard. Broadcasters, radio amateurs and a host of others use them.

Moving up further the very high frequency or VHF part of the spectrum is encountered. This contains a large number of mobile users. 'Radio taxis' and the like have allocations here, as do the familiar VHF FM broadcasts.

In the ultra high frequency or UHF part of the spectrum most of the terrestrial television stations are located. In addition to these there are more mobile users including the increasingly popular cellular telephones.

Above this in the super high frequency or SHF and extremely high frequency or EHF portions of the spectrum there are many uses for the radio spectrum. They are being used increasingly for commercial satellite and point to point communications.

How radio signals travel

Radio signals behave in a very similar way to light waves. Obviously there are some differences caused by the enormous variation in frequency between the two, but in essence they are the same.

A signal may be radiated or transmitted at a certain point, and the radio waves travel outwards, much like the waves seen on a pond if a stone is dropped into it. As they move outwards they become weaker as they have to cover a much wider area. However, they can still travel over enormous distances. Light can be seen from stars many light years away. Radio waves can also travel over similar distances. As distant galaxies and quasars emit radio signals these can be detected by radio telescopes which can pick up the minute signals and then analyse them to give us further clues about what exists in the outer extremities of the universe.

The direction in which radio waves travel can be changed. In just the same way that light can be reflected and refracted, so too can radio waves. This means that they do not just travel in exact straight lines and they can be made to travel over very long distances around the earth. It is because of this that signals on the short wave bands can be heard coming from places all over the world. To investigate why this occurs it is first necessary to look at the make-up of the atmosphere surrounding the earth.

Layers above the earth

The atmosphere above the Earth consists of many layers, as shown in Figure 2.5. Some of them have a considerable effect on radio waves whereas others do not. Closest to the Earth's surface is the troposphere. This region has very little effect on short wave frequencies below 30 MHz, although at frequencies above this it plays a major role. At certain times transmission distances may be increased from a few tens of kilometres to a few hundred kilometres. This is the area which governs the weather, and in view of this the weather and radio propagation at these frequencies are closely linked.

Above the troposphere, the stratosphere is to be found. This has little effect on radio waves, but above it in the mesosphere and thermosphere the levels of ionization rise in what is collectively called the ionosphere.

The ionosphere is formed as the result of a complicated process where the solar radiation together with solar and cosmic particles affect the atmosphere. This causes some of the air molecules to ionize, forming free electrons and positively charged ions. As the air in these areas is relatively sparse it takes

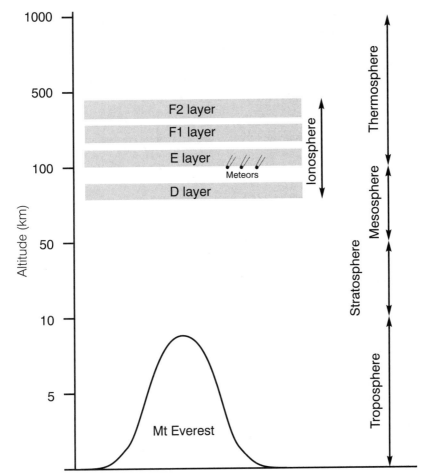

Figure 2.5 *Areas of the atmosphere*

some time for them to recombine. These free electrons affect radio waves causing them to be attenuated or bent back towards the earth.

The level of ionization starts to rise above altitudes of 30 km, but there are areas where the density is higher, giving the appearance of layers when viewed by their effect on radio waves. These layers have been designated the letters D, E, and F to identify them.

The degree of ionization varies with time, and is dependent upon the amount of radiation received from the sun. At night when the layers are hidden from the sun, the level of ionization falls. Some layers disappear while others are greatly reduced in intensity.

Other factors influence the level of ionization. One is the season of the year. In the same way that more heat is received from the sun in summer, so

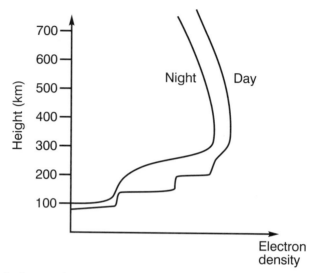

Figure 2.6 *Approximate ionization levels above the earth*

too the amount of radiation received by the upper atmosphere is increased. Similarly the amount of radiation received in winter is less.

The number of sunspots on the sun has a major effect on the ionosphere. These spots indicate areas of very high magnetic fields. It is found that the number of spots varies very considerably. They have been monitored for over 200 years and it has been found that the number varies over a cycle of approximately eleven years. This figure is an average, and any particular cycle may vary in length from about nine to thirteen years. At the peak of the cycle the sunspot number may reach 200 while at the minimum the number may be in single figures.

Under no circumstances should the sun be viewed directly, or even through dark sunglasses. This is very dangerous and can result in loss of eyesight.

Sunspots affect radio propagation because they emit vast amounts of radiation. In turn this increases the level of ionization in the ionosphere. Accordingly radio propagation varies in line with the sunspot cycle.

Each of the bands or layers in the ionosphere acts in a slightly different way, affecting different frequencies. The lowest layer is the D layer at a height of around 75 km. Instead of reflecting signals, this layer tends to absorb any signals which it affects. The reason for this is that the air density is very much greater at its altitude and power is absorbed when the electrons are excited. However, this layer only affects signals up to about 2 MHz or so. It is for this reason that only local ground wave signals are heard on the medium wave broadcast band during the day.

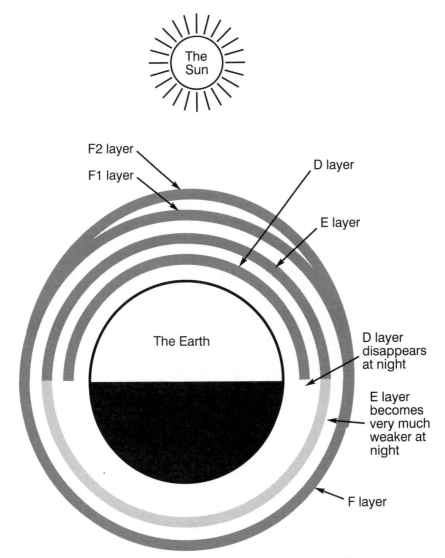

Figure 2.7 *Variation of the ionized layers during the day*

The D layer has a relatively low electron density and levels of ionization fall relatively quickly. As a result it is only present when radiation is being received from the sun. This means that it is much weaker in the evening and not present at night. When this happens it means that low frequency signals can be reflected by higher layers. This is why signals from much further afield can be heard on the medium wave band at night.

Above the D layer the E layer is found. At a height of around 110 km, this layer has a higher level of ionization than the D layer. It also reflects many of the signals which reach it, rather than absorbing them. However, there is a degree of attenuation with any signal reflected by the ionosphere. The atmosphere is still relatively dense at the altitude of the E layer. This means that the ions recombine quite quickly and levels of ionization sufficient to reflect radio waves are only present during the hours of daylight. After sunset the number of free ions falls relatively quickly to a level where they usually have little effect on radio waves.

The F layer is found at heights between 200 and 400 km. Like the E layer it tends to reflect signals which reach it. It has the highest level of ionization, being the most exposed to the sun's radiation. During the course of the day the level of ionization changes quite significantly. This results in it often splitting into two distinct layers during the day. The lower one called the F1 layer is found at a height of around 200 km, then at a height of between 300 and 400 km there is the F2 layer. At night when the F layer becomes a single layer its height is around 250 km. The levels of ionization fall as the night progresses, reaching a minimum around sunrise. At this time levels of ionization start to rise again.

Often it is easy to consider the ionosphere as a number of fixed layers. However, it should be remembered that it is not a perfect reflector. The various layers do not have defined boundaries and the overall state of the ionosphere is always changing. This means that it is not easy to state hard and fast rules for many of its attributes.

Ground wave

The signal can propagate over the reception area in a number of ways. The ground wave is the way by which signals in the long and medium wave bands are generally heard.

When a signal is transmitted from an antenna it spreads out, and can be picked up by receivers which are in the line of sight. Signals on frequencies in the long and medium wave bands (i.e. LF and MF bands) can be received over greater distances than this. This happens because the signals tend to follow the earth's curvature using what is termed the ground wave. It occurs because currents are induced in the earth's surface. This slows the wavefront down nearest to the earth causing it to tilt downwards, and enabling it to follow the curvature, travelling over distances which are well beyond the horizon.

The ground wave can only be used for signals below about 2 MHz. It is found that as the frequency increases the attenuation of the whole signal increases and the coverage is considerably reduced. Obviously the exact range will depend on many factors. Typically a high power medium wave station may be expected to be heard over distances of 100 miles. There are

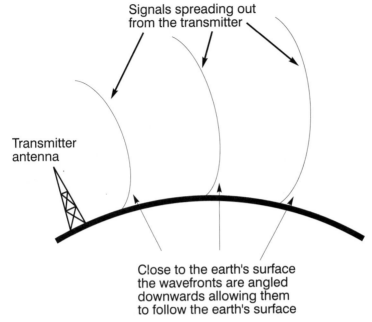

Signals spreading out
from the transmitter

Transmitter
antenna

Close to the earth's surface
the wavefronts are angled
downwards allowing them
to follow the earth's surface

Figure 2.8 *A ground wave*

also many low power broadcast stations running 100 watts or so. These might
have a coverage area extending to 15 or 20 miles.

As the effects of attenuation increase with frequency, even very high
power short wave stations are only heard over relatively short distances using
ground wave. Instead these stations use reflections from layers high up in the
atmosphere to achieve coverage to areas all over the world.

Skywaves

Radio signals travelling away from the earth's surface are called sky waves
and they reach the layers of the ionosphere. Here they may be absorbed,
reflected back to earth or they may pass straight through into outer space. If
they are reflected the signals will be heard over distances which are many
times the line of site. An exact explanation of the way in which the ionization
in the atmosphere affects radio waves is very complicated. However, it is
possible to gain an understanding of the basic concepts from a simpler
explanation.

Basically the radio waves enter the layer of increasing ionization, and as
they do so the ionization acts on the signal, bending it or refracting it back
towards the area of lesser ionization. To the observer it appears that the radio
wave has been reflected by the ionosphere.

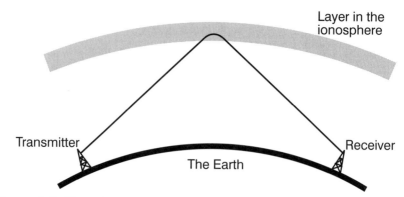

Figure 2.9 *Signals reflected and returned to earth by the ionosphere*

When the signal reaches the ionization, it sets the free electrons in motion and they act as if they formed millions of minute antennas. The electrons retransmit the signal, but with a slightly different phase. This has the result that the signal is made to bend away from the area of higher electron density. As the density of electrons increases as the signal enters the layer, the signal is bent back towards the earth's surface, so that it can often be received many thousands of kilometres away from where it was transmitted.

The effect is very dependent upon the electron density and the frequency. As frequencies increase much higher electron densities are required to give the same degree of refraction.

The way in which radio waves travel through the ionosphere, being absorbed, reflected or passing straight through, is dependent upon the frequency in use. Low frequency signals will be affected in totally different ways to those at the top end of the short wave spectrum. This is borne out by the fact that medium wave signals are heard over relatively short distances, and at higher frequencies signals from much further afield can be heard. It may also be found that on frequencies at the top end of the short wave spectrum, no signals may be heard on some days.

To explain how the effects change with frequency take the example of a low frequency signal transmitting in the medium wave band at a frequency of f_1. The signal spreads out in all directions along the earth's surface as a ground wave which is picked up over the service area. Some radiation also travels up to the ionosphere. However, because of the frequency in use the D layer absorbs the signal. At night the D layer disappears and the signals can then pass on being reflected by the higher layers.

Signals higher in frequency at f_2 pass straight through the D layer. When they reach the E layer they can be affected by it being reflected back to earth. The frequency at which signals start to penetrate the D layer in the day is

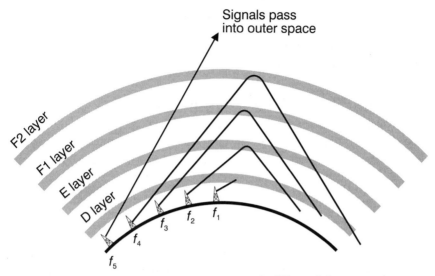

Figure 2.10 *Radio wave propagation at different frequencies*

difficult to define as it changes with a variety of factors including the level of ionization and angle of incidence. However, it is often in the region of 2 MHz or 3 MHz.

Also as the frequency increases the ground wave coverage decreases. Medium wave broadcast stations can be heard over distances of many tens of miles. At frequencies in the short wave bands this is much smaller. Above 10 MHz signals may only be heard over a few miles, dependent upon the power and antennas being used.

The E layer only tends to reflect signals in the lower part of the short wave spectrum to earth. As the frequency increases, signals penetrate further into the layer, eventually passing right through it. Once through this layer they travel on to the F layer. This may have split into two as the F1 and F2 layers. When the signals at a frequency of f_3 reach the first of the layers they are again reflected back to earth. Then as the frequency rises to f_4 they pass on to the F2 layer where they are reflected. As the frequency rises still further to f_5 the signals pass straight through all the layers, travelling on into outer space.

During the day at the peak in the sunspot cycle it is possible for signals as high as 50 MHz and more to be reflected by the ionosphere, but this figure falls to below 20 MHz at very low points in the cycle.

To achieve the longest distances it is best to use the highest layers. This is achieved by using a frequency which is high enough to pass through the lower layers. From this it can be seen that frequencies higher in the short wave spectrum tend to give the longer distance signals. Even so it is still possible for signals to travel from one side of the globe to the other on low frequencies at the right time of day. However, for this to happen good antennas are needed at the transmitter and receiver and high powers are generally required at the transmitter.

Distances and the angle of radiation

The distance which a signal travels if it is reflected by the ionosphere is dependent upon a number of factors. One is the height at which it is reflected, and in turn this is dependent upon the layer used for reflection. It is found that the maximum distance for a signal reflected by the E layer is about 2000 km, whereas the maximum for a signal reflected by the F layer is about 4000 km.

Signals leave the transmitting antenna at a variety of angles to the earth. This is known as the angle of radiation, and it is defined as the angle between the earth and the path the signal is taking.

It is found that those which have a higher angle of radiation and travel upwards more steeply cover a relatively small distance. Those which leave the antenna almost parallel to the earth travel a much greater distance before they reach the ionosphere, after which they return to the earth almost parallel to the surface. In this way these signals travel a much greater distance.

To illustrate the difference this makes, changing the angle of radiation from 0° to 20° reduces the distance for E layer signals from 2000 km to just 400 km. Similarly using the F layer, distances are reduced from 4000 km to 1000 km.

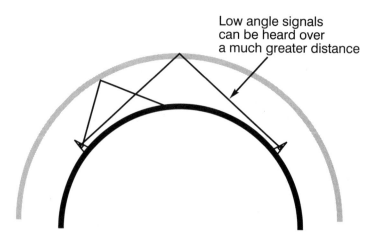

Low angle signals can be heard over a much greater distance

Figure 2.11 *Effect of the angle of radiation on the distance achieved*

For signals which need to travel the maximum distance this shows that it is imperative to have a low angle of radiation. However, broadcast stations often need to make their antennas directive to ensure the signal reaches the correct area. Not only do they ensure they are radiated with the correct azimuth, they also ensure they have the correct angle of elevation or radiation so that they are beamed to the correct area. This is achieved by altering the antenna parameters.

Multiple reflections

The maximum distance for a signal reflected from the F2 layer is about 4000 km. However, radio waves travel much greater distances than this around the world. This cannot be achieved using a single reflection, but instead several are used, as shown in Figure 2.12.

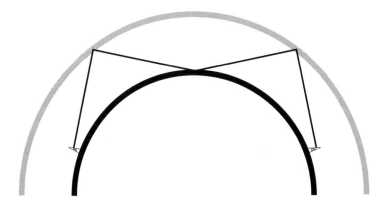

Figure 2.12 *Several reflections used to give greater distances*

To achieve this the signals travel to the ionosphere and are reflected back to earth in the normal way. Here they can be picked up by a receiver. However, the earth also acts as a reflector because it is conductive. In fact it is found that areas which are more conductive act as better reflectors. Not surprisingly the sea acts as an excellent reflector. Once reflected at the earth's surface the signals travel towards the ionosphere where they are again reflected back to earth.

At each reflection the signal suffers some attenuation. This means that it is best to use a path which gives the minimum number of reflections as shown in Figure 2.13. Lower frequencies are more likely to use the E layer and as the maximum distance for each reflection is less, it is likely to give lower signal strengths than a higher frequency using the F layer to give less reflections.

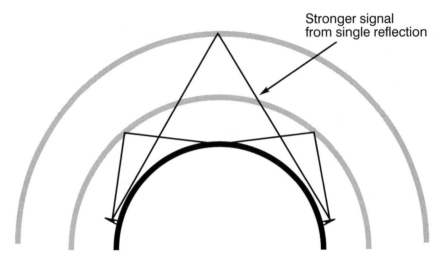

Figure 2.13 *The minimum number of reflections usually gives the best signal*

Chordal hop

Not all reflections around the earth occur in exactly the ways described. It is possible to calculate the path which would be taken, the number of reflections, and hence the path loss and signal strength expected. Sometimes signal strengths appear higher than would be expected. In conditions like these it is likely that a propagation mode called chordal hop is being experienced.

When this happens it is found that the signal travels to the ionosphere where it is reflected, but instead of returning to the earth it takes a path which

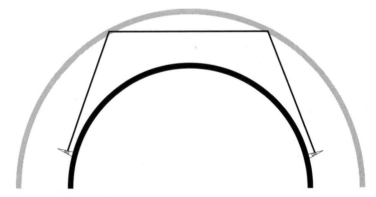

Figure 2.14 *Chordal hop propagation*

intersects with the ionosphere again, only then being reflected back to earth.

It can be seen from the way in which this occurs that fewer reflections are needed to cover a given distance. As a result signal strengths are higher when this mode of propagation is used.

Critical frequency

When a signal reaches a layer in the ionosphere it undergoes refraction and often it will be reflected back to earth. The steeper the angle at which the signal hits the layer the greater the degree of refraction is required. If a signal is sent directly upwards this is known as vertical incidence, as shown in Figure 2.15.

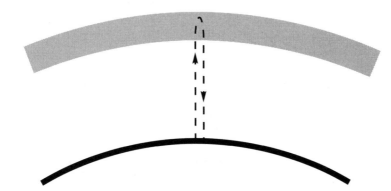

Figure 2.15 *Vertical incidence*

For vertical incidence there is a maximum frequency for which the signals will be returned to earth. This frequency is known as the critical frequency. Any frequencies higher than this will penetrate the layer and pass right through it onto the next layer or into outer space.

MUF

When a signal is transmitted over a long distance path it penetrates further into the reflecting layer as the frequency increases. Eventually it passes straight through. This means that for a given path there is a maximum frequency which can be used. This is known as the maximum usable frequency or MUF. Generally the MUF is three to five times the critical frequency depending upon which layer is being used and the angle of incidence.

For optimum operation a frequency about 20% below the MUF is normally used. It is also found that the MUF varies greatly depending upon the state of the ionosphere. Accordingly it changes with the time of day, season, position in the eleven year sunspot cycle, and the general state of the ionosphere.

LUF

When the frequency of a signal is reduced, further reflections are required and losses increase. As a result there is a frequency below which the signal cannot be heard. This is known as the lowest usable frequency or LUF.

Skip zone

When a signal travels towards the ionosphere and is reflected back towards the earth, the distance over which it travels is called the skip distance, as shown in Figure 2.16. It is also found that there is an area over which the

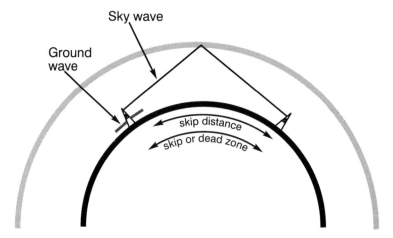

Figure 2.16 *Skip zone and skip distance*

signal cannot be heard. This occurs between the position where the signals start to return to earth and where the ground wave cannot be heard. The area where no signal is heard is called the skip or dead zone.

State of the ionosphere

Radio propagation conditions are of great importance to a vast number of users of the short wave bands. Broadcasters, for example, are very interested in them as are other professional users. To detect the state of the ionosphere, an instrument called an ionosonde is used. This is basically a form of radar system which transmits pulses of energy up into the ionosphere. The reflections are then monitored and from them the height of the various layers can be judged. Also by varying the frequency of the pulses, the critical frequencies of the various layers can be judged.

Fading

One of the characteristics of listening to short wave stations is that some signals appear to fade in and out all the time. These alterations are taken as a matter of course by listeners who are generally very tolerant to the imperfections in the quality of the signal received from the ionosphere. There are a number of different causes for fading but they all hinge around the fact that the state of the ionosphere is constantly changing.

The most common cause of fading occurs as a result of multipath interference. This occurs because the signal leaves the antenna at a variety of different angles and reaches the ionosphere over a wide area. As the ionosphere is very irregular the signal takes a number of different paths as shown in Figure 2.17. The changes in the ionosphere cause the lengths of these different paths to vary. This means that when the signals come together at the receiving antenna they pass in and out of phase with one another. Sometimes they reinforce one another; at other times they cancel each other out. This results in the signal level changing significantly over periods of even a few minutes.

Another reason for signal fading arises out of changes in polarization. It is found that when the ionosphere reflects signals back to earth they can be in any polarization. For the best reception, signals should have the same polarization as the receiving antenna. As the polarization of the reflected

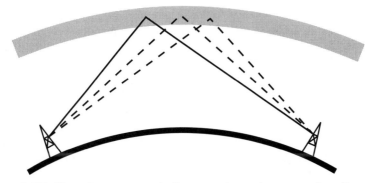

Figure 2.17 *Signals can reach the receiver via several paths*

wave will change dependent upon the ionosphere, the signal strength will vary according to the variations in polarization.

In some instances the receiver may be on the edge of the skip zone for a particular signal. When this happens any slight variations in the state of the ionosphere will cause the receiver to pass into or out of the skip zone, giving rise to signal strength variations.

On other occasions severe distortion can be heard particularly on amplitude modulated signals. This can occur when different sideband frequencies are affected differently by the ionosphere. This is called selective fading and it is often heard most distinctly when signals from the ground and sky waves are heard together.

Ionospheric disturbances

At certain times ionospheric propagation can be disrupted and signals on the short wave bands can completely disappear. These result from disturbances on the sun called solar flares. These flares are more common at times of high sunspot activity, but they can occur at any time.

When a flare occurs there is an increase in the amount of radiation which is emitted. The radiation reaches the earth in about eight minutes and causes what is termed a sudden ionospheric disturbance (SID). This is a fast increase in the level of absorption in the D layer lasting anywhere from a few minutes to a few hours. This can affect all or part of the short wave spectrum, dependent upon the level of increase in radiation.

The next stage of the process sees changes in the solar wind. Under normal conditions there is a flow of particles away from the sun. This is the solar wind, and the earth's magnetic field is able to give protection against this. However, after a flare there is a considerable increase in the solar wind. This normally occurs about 20 to 30 hours after the flare. When it arrives it starts a complicated chain of events. Large variations in the earth's magnetic field can be observed and a visible aurora may be seen in locations towards the poles. Generally it is necessary to be at a latitude of greater than about 55° to see this. Although the short wave bands may initially improve after the SID, the increase in solar wind causes a major degradation in communications over the HF portion of the spectrum. This mainly results from a drastic decrease in the level of ionization in the ionosphere. As a result radio signals are not reflected back to earth in the usual way, causing a radio blackout.

During some stages of the aurora very high levels of ionization are seen towards the poles. As a result signals may be reflected back to earth in these regions at frequencies up to about 150 MHz. When this happens signals generally have a distinctive buzz superimposed upon them. This results from the constantly changing nature of the ionosphere under these conditions.

This blackout in radio communications may last anywhere from a few hours to a few days after which the bands slowly recover. The first signs of the end of the blackout are normally seen at the low end of the spectrum first. It is also found that further disruption may occur after 28 days, the period of the sun's rotation.

Propagation prediction

While it is possible to gain a feel for how the various areas of the radio spectrum behave, predictions for the radio conditions are published, or broadcast over the air. These predictions can be very valuable for the listener, enabling him or her to know when to listen.

It is not possible to predict exactly what the conditions will be like. Instead they are an indication of what is expected, and they should only be treated as a probability, giving the chances that a particular path or route on a frequency will be open for radio communications.

A comparison of the perceived conditions against the predictions is also dependent upon the antenna in use, and the exact location. Most predictions indicating the performance of a particular path will relate to the capital of the country. In other words any predictions for the UK will normally relate to a station located in London. Also the antenna in use will have a marked effect. A poor antenna will reduce the chance of hearing stations from a given area.

A number of transmitting stations carry predictions about the ionosphere. Of these the most famous is WWV, an American standard frequency transmission found on 2.500 MHz, 5.000 MHz, 10.000 MHz and 20.000 MHz. The station is primarily used as a highly accurate frequency transmission, but it also transmits bulletins giving the solar flux, A index for the previous day, the K index of the condition of the earth's magnetic field.

The solar flux is a measure of the sun's radiation at a frequency of 2800 MHz. A figure of 60 indicates that conditions will be quiet, but as it rises it gives an indication of an improvement. Readings of 80 and higher mean that the higher frequency bands above 20 MHz or so will be open for signals. Flux readings around 200 occur towards the peak of the cycle and give an indication that frequencies even above 30 MHz can support long distance communication via the F2 layer.

The A and K indices give a measure of the earth's magnetic field. Low values of these figures give an indication of better HF conditions. An A index of 10 or less is indicative of good HF conditions, but the figure can rise to over 100, but this is not common.

The K index is similar to the A index, but is more up to date. A figure of between zero and one indicates quiet geomagnetic field conditions and the likelihood of good HF conditions. One to three indicate unsettled conditions and four and above show high levels of geomagnetic activity and poor HF conditions.

VHF and above

At frequencies above the limit of ionospheric propagation but below about 3000 MHz communication can be established over distances greater than the ordinary line of sight. This is as a result of effects within the troposphere. As most of the conditions which govern our weather occur in the troposphere, there are usually many links between the weather and radio propagation conditions at these frequencies.

Under normal conditions signals at these frequencies travel more than the line of sight. Prior to the 1940s it was generally thought that communication over distances greater than the line of sight was not possible, but experience soon showed this was not true. As a very rough guide it is usually possible to achieve distances at least a third greater than this. This is possible because of the varying refractive index of the air above the earth's surface. An increase in the pressure and humidity levels close to the earth's surface mean that the refractive index is greater than the air higher up. Like light waves radio waves can be refracted, and they bend towards the area of the greater refractive index. This means that the signals tend to follow the curvature of the earth and travel over distances which are greater than just the line of sight. An additional effect is that of diffraction where the signal diffracts around the earth's curvature.

Greater distances

At times signals can be received over much greater distances than 4/3 of the line of sight. At times like these terrestrial television channels may be subject to interference, although radio amateurs relish these conditions as they give the possibility of contacts over much greater distances than normal.

There are a variety of mechanisms by which signals can be propagated over these greater distances. Usually it is possible to predict when there is a likelihood of them occurring as there is a strong correlation between them and certain weather conditions. Usually the extended distances result from the normal gradient in refractive index becoming much steeper. In this way the degree of bending is increased, allowing the signals to follow the curvature of the earth for greater distances.

A number of weather conditions may cause this increase. A high pressure area may cause the conditions which can increase the normal propagation distance. A high pressure is normally associated with warm weather, especially in summer. Under these conditions the hot air rises and cold air comes in to replace it. This accentuates the density gradient normally present and the change in refractive index occurring as a result of this can be very sharp.

Other weather conditions can also bring about similar increases in the change of refractive index. Cold weather fronts can have the same effect.

Here a mass of warm air and a mass of cold air meet. When this occurs the warm air rises above the cold air bringing about similar conditions. Cold weather fronts normally move more quickly than high pressure areas, and as a result an increase in propagation distance due to cold fronts is normally more short lived than those caused by high pressure areas.

Other local conditions may give rise to increases in propagation distance. Convection in coastal areas in warm weather, the rapid cooling of the earth and the air closest to it after a hot day, or during frosty weather. Subsidence of cool, moist air into valleys on calm summer evenings can give rise to these changes.

Sometimes the changes in refractive index can trap the signals between two layers forming a type of duct or waveguide. When this happens signals may be carried for several hundreds of kilometres.

It is found that tropospheric bending and ducting is experienced more at higher frequencies. Its effects are comparatively small at frequencies at the top of the HF portion of the spectrum, and increases steadily into the VHF and UHF portion of the spectrum. At higher frequencies the effects are still noticed, but other factors start to limit the range.

Tropo-scatter

The effects of tropospheric bending are very dependent upon the weather. This is shown by the fact that television signals in the UHF band are only occasionally affected by interference from distant signals. As such it is not possible to rely on these modes for extending the range of a communications link. Where links are required a mode of propagation known as tropo-scatter can be used. This form of propagation relies on the fact that within the troposphere there are masses of air with a slightly different refractive index

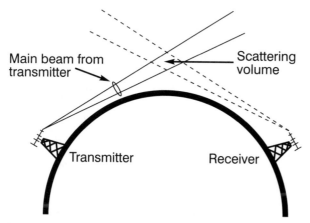

Figure 2.18 *The mechanism behind tropo-scatter*

which are moving around randomly. These arise because of the continually moving nature of air, and the differing temperatures of different parts.

These masses of air reflect and bend the signals, and small amounts of the signal are returned to earth. In view of the small amounts of signal which are returned to earth using this mode, high transmitter powers, high gain antennas and sensitive receivers are required. Nevertheless this form of propagation can be very useful for links over distances around 1000 to 1500 km.

Sporadic E

Sometimes in summer it is possible for signals to be audible in the bands at the top end of the short wave spectrum at the bottom of the sunspot cycle. When the maximum usable frequency may be well below the frequencies in question signals from stations up to 2000 km distance may be heard in summer. This occurs as a result of a form of propagation known as sporadic E.

Sporadic E used to be well known when television transmissions used frequencies around 50 MHz, and sometimes in summer reception would be disturbed by interference from distant stations. Even today reception of VHF FM signals can be disturbed when frequencies around 100 MHz are affected by it. The maximum frequencies it generally affects are up to around 150 MHz, although it has affected higher frequencies up to about 220 MHz on very rare occasions.

Sporadic E occurs as a result of highly ionized areas or clouds forming within the E layer. These clouds have a very patchy structure and may measure anywhere between 100 km and 1000 km across and less than 2 km thick. This means that propagation is quite selective when the clouds are small with signals coming from a particular area. However, their electron density is much greater than that normally found in the E layer and as a result signals with much higher frequencies are reflected. It is also found that there are irregularities in the structure of the clouds and this makes them opaque to lower frequency signals.

At the onset of propagation via sporadic E the level of ionization starts to build up. At first only the less high frequencies are affected. Those at the top end of the HF part of the spectrum are affected first. As the levels of ionization increase further, frequencies into the VHF region are reflected.

Sporadic E normally occurs in the summer, reaching a peak broadly around mid summer. Even so frequencies at the top of the short wave band may be affected on some days at least a couple of months either side of this. Frequencies well up into the VHF portion of the spectrum are normally affected closer to the centre of the season because much higher ionization levels are required. It is also found that the very high frequencies are not affected for as long. Sometimes signals may only be heard for a few minutes before propagation is no longer supported.

The sporadic nature of this form of propagation means that it is very difficult to predict when it will occur. Even when propagation is supported by this mode it is very variable. The ionized clouds are blown about in the upper atmosphere by the swiftly moving air currents. This means that the area from which stations are heard can change. Accordingly sporadic E is not a mode normally used for commercial communications. However, radio amateurs make use of it to make long distance communications on bands like 2 metres at 144 MHz.

Meteor scatter

Meteor scatter or meteor burst communication is a useful form of propagation for distances of up to about 2000 km. It is generally used for data links and for applications where real time communications are not required, for which it provides a cost effective method of communication.

Meteor scatter relies upon the fact that meteors are constantly entering the earth's atmosphere. It is estimated that every day about 75 000 000 enter. The vast majority of them are small, and do not produce the characteristic visible trail in the sky. In fact most meteors are only about the size of a grain of sand and any which are an inch across are considered to be large.

The meteors enter the atmosphere at speeds of up to 75 km/second and as the atmosphere becomes more dense they burn up, usually at heights of around 80 km. The heat generated from the friction with the air causes the atoms on the surface of the meteor to vaporize. The atoms become ionized and, in view of the speeds, leave a trail of positively charged ions and negatively charged atoms.

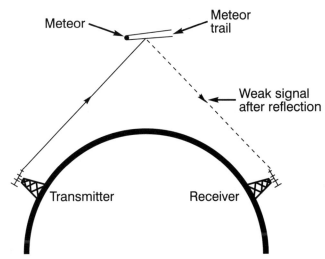

Figure 2.19 *Meteor scatter link*

The trails do not normally last very long. As the density of the air is relatively high, the electrons and ions are able to recombine quickly. As a result the trails normally only last for a second or so. However, the level of ionization is very high and they are able to reflect radio waves up to frequencies of 100 MHz and more. While the level of ionization is very high, the area which can reflect signals is very small, and only a small amount of energy is reflected. Despite this there is just enough for a sensitive receiver to pick up.

Many meteors come from the sun, and there are two main types. Most enter the atmosphere on a random basis, while others are from meteor showers. The showers occur at specific times of the year and occur as the earth passes through areas around the sun where there is a large amount of debris. When these showers are viewed at night it will be seen that the meteors appear to come from a specific point in the sky. The showers are named after the point in the sky from where they appear to come, as shown in Table 2.1. This is known as the radiant.

A wide range of frequencies can support meteor scatter communications, although at lower frequencies signals suffer from attenuation in the D layer of the ionosphere. Also if frequencies in the HF portion of the spectrum are

Table 2.1 Major meteor showers

Shower	*Begins*	*Maximum*	*Ends*
Quadrantids	1 January	3 January	6 January
April Lyrids	19 April	21 April	24 April
Eta Aquarids	1 May	4 May	7 May
June Lyrids	10 June	15 June	21 June
Ophiuchids	17 June	20 June	26 June
Capricornids	10 July	26 July	15 August
Delta Aquarids	15 July	27 July	15 August
Pisces Australids	15 July	30 July	20 August
Alpha Capricornids	15 July	2 August	25 August
Iota Aquarids	15 July	6 August	25 August
Perseids	25 July	12 August	18 August
Orionids	16 October	21 October	26 October
Taurids	20 October	4 November	25 November
Cepheids	7 November	9 November	11 November
Leonids	15 November	17 November	19 November
Geminids	7 December	14 December	15 December
Ursids	17 December	22 December	24 December

used then there is the possibility of propagation by reflection from the ionosphere. These two reasons mean that meteor scatter operation is generally confined to frequencies above 30 MHz. Operating above these frequencies has the further advantage that both galactic and artificial noise are less – a vital factor when considering the low signal levels involved in this mode of communication.

Generally most meteor scatter operation takes place between 40 and 50 MHz, although there is some between 30 and 40 MHz. The top limit is governed more by the fact that television transmissions previously occupied frequencies above 50 MHz, and still do in some countries. In addition to this reflections from the meteor trails start to deteriorate above this limit, although radio amateurs are still able to use them up to their 144 MHz band.

Frequencies above 3 GHz

At frequencies above about 3 GHz, the distances which can be achieved are not normally much in excess of the line of sight. This means that if greater distances are to be achieved, antennas must be placed higher above the earth's surface to increase the distance of the horizon.

Other effects are also noticed. Signals are absorbed more by atmospheric conditions. Water vapour and rain can give significant degrees of attenuation making the signal dependent on many of the conditions which limit normal visibility.

3 Capacitors, inductors and filters

There is a wide variety of components used in radio circuits. Most of them are commonly used in the more standard circuits, but others are used either in slightly different ways or they are of greater importance in radio circuitry. Inductors, capacitors and filters are of great importance in a number of radio circuits. These will be explained in this chapter. Components including resistors, semiconductors, etc., which are widely used in other areas of electronics, will not be addressed here. It is assumed that the reader has a basic understanding of them from other areas of electronics.

Capacitors

One of the most basic components found in a radio frequency circuit is the capacitor. It has the ability to store electric charge, and in its simplest form it just consists of two parallel plates as shown in Figure 3.1. When a battery or other voltage source is connected across the plates, one plate has an excess of electrons while the other has too few. In this way one plate becomes positively charged, while the other is negatively charged. The positive charge on one plate is equal and opposite to the negative charge on the other. If the battery is removed the capacitor will retain its charge. However, if a resistor is placed across the plates, a current will flow until the capacitor becomes discharged.

As the charge on the capacitor falls, so the voltage across the plates is reduced. This means that the current will be reduced, and in turn the rate at which the charge is reduced falls. Therefore the voltage across the capacitor falls in an exponential fashion, gradually approaching zero, as shown in Figure 3.2.

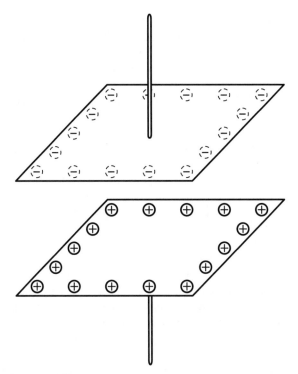

Figure 3.1 *A capacitor*

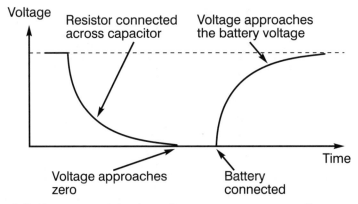

Figure 3.2 *Decay and rise in voltage across a capacitor*

Similarly, when a battery in series with a resistor is connected across the capacitor the rate at which the voltage rises is much faster at first than after it has been charging for some while. Eventually it reaches a point when it is virtually fully charged and almost no current flows. In theory the capacitor never becomes fully charged as the curve is asymptotic. However, in reality it reaches a point where it can be considered to be fully charged or discharged and no current flows. The rate at which the voltage rises or decays is dependent upon the resistance in the circuit. The greater the resistance the smaller the amount of charge which is transferred and the longer it takes for the capacitor to charge or discharge.

So far consideration has been given to the case where a battery has been connected to charge the capacitor and disconnected and a resistor applied to charge it up. If an alternating waveform, which by its nature is continually changing, is applied to the capacitor, then it will be in a continual state of charging and discharging. For this to happen a current must be flowing in the circuit. In this way a capacitor will allow an alternating current to flow, but it will block a direct current. As such, capacitors are used for coupling an AC signal between two circuits which are at different steady state potentials.

In an electric circuit it is found that the voltage and current are not exactly in phase. Because current flows through the capacitor when there is a change in voltage, the current leads the voltage by 90°. The maximum rate of change in voltage takes place when the voltage is midway between the two peaks. This is when the maximum current flows. The minimum rate of change of voltage occurs at either peak and hence the current is at a minimum.

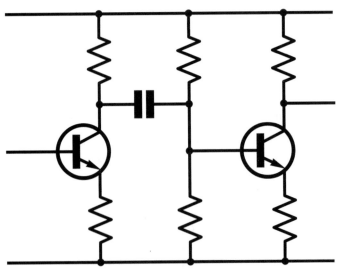

Figure 3.3 *An example of a circuit where a capacitor is used to block DC but pass the AC*

Capacitors can also be used to store charge in a smoothing circuit. They are most obviously used for this purpose in power supply circuits. Here an incoming alternating supply is rectified leaving a signal which is still varying from zero to the peak of the AC waveform. This is of no use for powering other electronic circuits and accordingly smoothing is required. Initially the capacitor charges up as the waveform voltage increases, and as the input voltage falls the capacitor starts to discharge and in doing so supplies current to the circuitry. In successive cycles the capacitor only charges up when the input voltage exceeds that of the capacitor voltage. This takes place during a short part of the waveform, as shown in Figure 3.4.

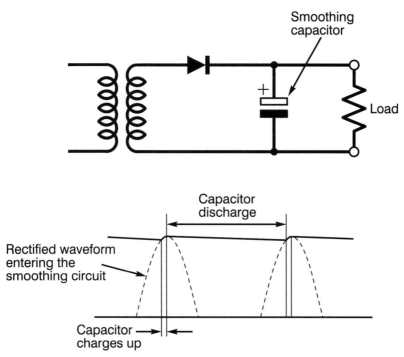

Figure 3.4 *Capacitor used for smoothing*

Capacitance is measured in farads. A capacitor is said to have a capacitance of one farad if a charge of one coulomb (i.e. a current of one amp flowing for one second) sets up a potential difference of one volt across it. A one farad capacitor would be physically very large, and usually capacitors with a much smaller capacity are required. As a result most capacitors are measured in microfarads (μF) which are millionths of a farad, nanofarads (nF) which are thousand-millionths of a farad, or picofarads (pF) which are million-millionths of a farad. Typically those required in audio frequency circuits and power supplies are of the order of a few microfarads to a few hundreds or thousands of microfarads. Those used in radio frequency circuits are usually measured in picofarads or nanofarads.

In a direct current circuit, where there may be a battery and a resistor, it is the resistor that resists the flow of current in the circuit. The same is true for an alternating current circuit with a capacitor. A capacitor with a small plate area will only be able to store a small amount of charge, and this will impede the flow of current. A larger capacitor will allow a greater flow of current. A capacitor is said to have a certain reactance. This name is chosen to be different to that of a resistor, but it is measured in ohms just the same. The reactance of a capacitor is dependent upon the value of the capacitor and also the frequency of operation. The higher the frequency the smaller the reactance. The actual reactance can be calculated from the formula:

$$X_C = \frac{1}{2\pi f C}$$

where X_C is the capacitive reactance in ohms
    f is the frequency in hertz
    C is the capacitance in farads

This reactance is measured in ohms, and the current flowing in a circuit can be calculated from Ohm's law in the same way as for a resistor.

A number of capacitors may be used in parallel or series. In these cases it is often necessary to be able to calculate the capacitance of the components together. If they are placed in parallel this is effectively the same as increasing the size of the capacitor plates, and the total capacitance is simply the sum of all the individual components.

$$C_{TOTAL} = C_1 + C_2 + C_3 + \ldots$$

If the capacitors are placed in series the equation is a little more complicated:

$$\frac{1}{C_{TOTAL}} = \frac{1}{C_1} + \frac{1}{C_2} + \frac{1}{C_3} + \ldots$$

If only two capacitors are used this simplifies to:

$$C_{TOTAL} = \frac{C_1 \times C_2}{C_1 + C_2}$$

If a capacitor and resistor are in series it is not possible to add the resistance to the reactance to gain the total value of the impedance. This must

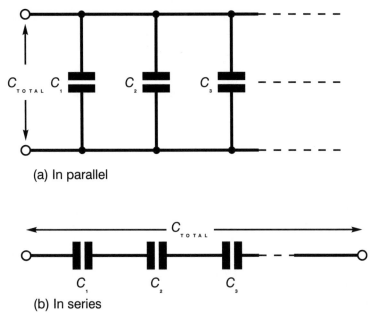

(a) In parallel

(b) In series

Figure 3.5 *Capacitors in parallel and series*

be done vectorially because of the 90° phase difference between the two and it is calculated as below:

$$Z_{\text{total}} = \sqrt{X_c^2 + R^2}$$

where Z_{total} is the total impedance

Inductors

Inductors are another vital component in radio frequency circuits, and have a number of similarities. Instead of storing electric charge, inductors store energy in the magnetic field. When current flows in the inductor, a magnetic field is built up.

If a steady voltage is applied to an inductor the current flowing though it would be governed by the resistance of the wire. However, if an alternating voltage is applied the current is no longer just related to the resistance of the coil. In most cases the resistance of the coil is small and can be ignored. Instead the major effect is the inductive reactance. This is proportional to the inductance of the coil and the frequency. This inductance is measured in henries.

An inductor is said to have an inductance of one henry if a current changing at a rate of one amp per second sets up a potential difference of one volt across it. This assumes that the coil has no resistance. Like the capacitor the basic unit is often too large for most radio frequency applications. As a result most inductors for practical applications have values measured in millihenries (thousandths of a henry) or microhenries (millionths of a henry).

For any coil the inductance can be calculated. It is simply:

$$X_L = 2\pi f L$$

where X_L is the inductive reactance in ohms
f is the frequency in hertz
L is the inductance in ohms

In the same way that the current flowing through a capacitor could be calculated using Ohm's law and a knowledge of the reactance and voltage, the same is true for inductors from a knowledge of the inductive reactance and voltage.

When combinations of inductors are put together the total inductance can be calculated. If they are connected in parallel the following formula must be used:

$$\frac{1}{L_{TOTAL}} = \frac{1}{L_1} + \frac{1}{L_2} + \frac{1}{L_3} + \ldots$$

If there are only two inductors this simplifies to:

$$R_{TOTAL} = \frac{L_1 \times L_2}{L_1 + L_2}$$

If the inductors are placed in series the values are simply added:

$$L_{TOTAL} = L_1 + L_2 + L_3 + \ldots$$

If an inductor and resistor are connected in series it is not possible to add the resistance to the reactance to gain the total value of the impedance. This must be done vectorially because of the 90° phase difference between the two and it is calculated as below:

$$Z_{total} = \sqrt{X_L^2 + R^2}$$

where Z_{total} is the total impedance

When an alternating voltage is applied to an inductor, a back EMF is generated and this results in the current waveform lagging behind the voltage by 90°. It is for this reason that a reactance and resistance cannot be directly added together.

Tuned circuits

One of the most important aspects associated with inductors and capacitors is that they can be combined together to form a tuned circuit. A mechanical equivalent of an electrical tuned circuit is a pendulum. This moves backwards and forwards at an exact frequency defined by the length and weight of the pendulum. Once the pendulum has been set in motion, very little energy is lost and the pendulum continues to move to and fro for some while. The losses are mainly caused by the air resistance, and any friction at the mount for the pendulum at the top. Despite the loss of energy the pendulum still has the same frequency.

The electrical equivalent is formed when an inductor and capacitor are connected together. It has already been mentioned that the current leads the voltage for a capacitor and lags behind it for an inductor. At a certain frequency the magnitude of the reactances of the inductor and capacitor will be the same. As the current waveforms lead the voltage in one case and lag in the other, they are 180° out of phase with one another. At the point where the reactances are the same they become equal and opposite and cancel one another out, leaving only the resistance within the circuit. As this is normally relatively small, it means that the impedance of the circuit changes by a large amount over a relatively small frequency difference.

The way in which the capacitor and an inductor are connected governs how the impedance changes. If they are in series, they create what is called a series tuned circuit, as shown in Figure 3.6. The impedance of this circuit changes from being very high away from the resonant frequency to very low at resonance.

If the capacitor and inductor are connected in parallel the impedance away from resonance is low. At resonance this rises to a maximum as shown. This

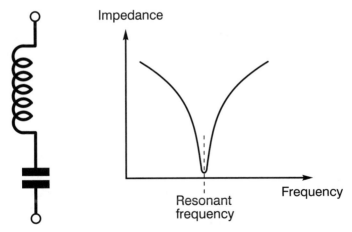

Figure 3.6 *A series tuned circuit*

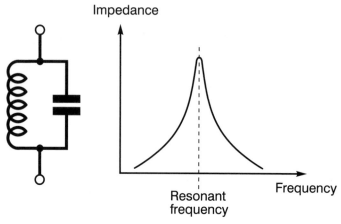

Figure 3.7 *A parallel tuned circuit*

is called a parallel resonant circuit. It is used more widely than the series tuned circuit and it can be seen in circuits from crystal sets to modern radio receivers and transmitters.

Mutual inductance and transformers

Apart from ordinary inductors, transformers are widely used in radio frequency circuits. It is found that when two windings are placed close to one another the magnetic field generated by one affects the second and vice versa. This means that an alternating or changing waveform in one will generate a changing field. This in turn will induce an electromotive force in the second and a voltage will be seen at the output. The winding at which the power enters the transformer is called the primary, whereas the one from which the power leaves is called the secondary.

It is found that the ratio of the turns on the primary and secondary windings is related to the input and output voltages. The voltage ratio between the input and output is proportional to the turns ratio.

$$\frac{V_{in}}{V_{out}} = \frac{N_p}{N_s}$$

where V_{in} is the input voltage, i.e. the voltage across the primary

V_{out} is the output voltage, i.e. the voltage across the secondary

N_p is the number of turns in the primary

N_s is the number of turns in the secondary

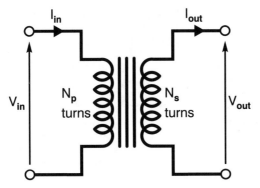

Figure 3.8 *A transformer*

The current ratio is the inverse of the voltage ratio. In other words the greater the number of turns at the output, the higher the voltage but the lower the current when compared to the input or primary.

$$\frac{I_{\text{in}}}{I_{\text{out}}} = \frac{N_s}{N_p}$$

where I_{in} is the input current, i.e. the current flowing in the primary

$\quad\quad I_{\text{out}}$ is the output current, i.e. the current flowing in the secondary

It can be seen that the voltage and current can be stepped up or down. However, there is no power change in an ideal transformer. For example, if the voltage is doubled the current is halved. In practice every transformer has some loss which is dissipated as heat, but this is normally small as most transformers have a high degree of efficiency.

Transformers have a variety of uses. These are often used to provide isolation between two circuits. This is the case for a mains transformer where the transformer isolates the electronics in a radio set, for example from the mains. It may also be used to give the required voltage, combining two functions in the same component.

In terms of radio frequency design, transformers can be used to provide impedance matching. Take the example of a transformer which has a low number of turns on the input and a high number at the output. The voltage current ratio at the input is lower than that at the output. Using Ohm's law it can be seen that the impedance at the input is lower than the output. This can be calculated from a knowledge of the turns ratio.

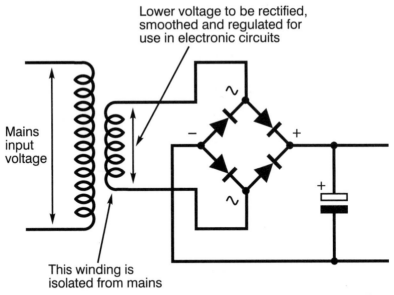

Lower voltage to be rectified, smoothed and regulated for use in electronic circuits

Mains input voltage

This winding is isolated from mains

Figure 3.9 *A transformer used in a mains power supply circuit*

$$\frac{Z_{in}}{Z_{out}} = \frac{N_p^{\,2}}{N_s^{\,2}}$$

where Z_{in} is the impedance in the primary circuit

Z_{out} is the impedance in the secondary circuit

Transformers find a number of uses for impedance matching. For example, they may be used in antenna systems when different impedances need to be matched. They may also be used in radio circuitry to ensure the optimum signal transfer between stages which have different impedances.

Transformers may also be tuned. In this way they can combine the aspects of isolating the DC component of one circuit with another, with the transforming any impedances as required, and acting as a tuned circuit. In some cases only the primary of the transformer is tuned. If both are tuned it is common for them to be set to slightly different frequencies to give the required bandwidth.

Q factor

The performance of a tuned circuit is very important in many applications. The width over which the circuit will cause signals to be accepted, and the losses which it introduces, are of great importance because they determine

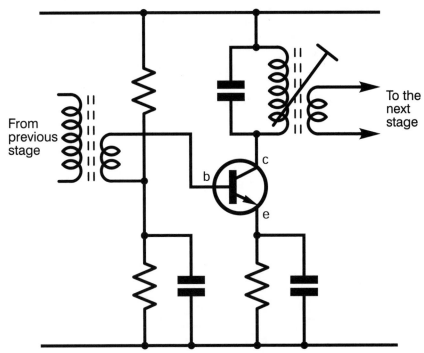

Figure 3.10 *A tuned transformer used in the filter stages of a radio receiver*

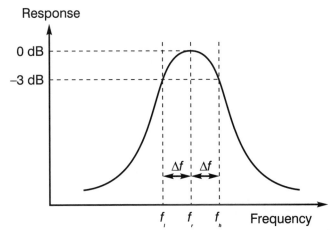

Figure 3.11 *Response curve of a parallel tuned circuit*

the performance of receivers, transmitters and a variety of other pieces of equipment. To define the performance of a tuned circuit or inductor a factor known as the quality or Q factor is used.

A typical response curve for a tuned circuit is shown in Figure 3.11. The case of a parallel tuned circuit is taken here, but the concept of Q is equally applicable to a series tuned circuit. The first point to define is the bandwidth. For the purposes of this calculation the point where the response has fallen to $1/\sqrt{2}$ or 0.707 is used. This is also the point where the power has fallen by half, i.e. $-3\,\text{dB}$.

The relationship between the resonant frequency and the bandwidth can then be calculated:

$$Q = \frac{f_r}{f_h - f_l}$$

$$= \frac{f_r}{2\Delta f}$$

where f_r is the resonant or centre frequency
f_h is the $-3\,\text{dB}$ point above resonance
f_l is the $-3\,\text{dB}$ point below resonance
Δf is the frequency from the resonant frequency to the $-3\,\text{dB}$ point

A high Q circuit is able to perform its functions of selectivity far more effectively because circuit losses are much lower. This can be seen in the case of a series tuned circuit where any resistance present will develop a voltage across it and mean that even after the reactance from the inductor and capacitor have been cancelled out there is still some resistance there. In the case of a parallel tuned circuit the resistance losses damp down the resonance as shown in Figure 3.12. Here the high Q circuit has a very sharp curve, whereas the low Q one has a much shallower curve with the point of resonance much less well defined.

It can be seen that the DC resistance of a circuit has a direct bearing on the Q. In most cases the majority of the resistance is in the inductor, as losses in capacitors are very much smaller. This means that the Q of a tuned circuit is virtually that of the inductor alone. It is possible to calculate the Q from a knowledge of the resistance because the Q of the circuit is the ratio of the inductive reactance to the resistance in the circuit:

$$Q = \frac{2\pi L}{R}$$

where Q is the quality factor
L is the inductance
R is the resistance of the circuit

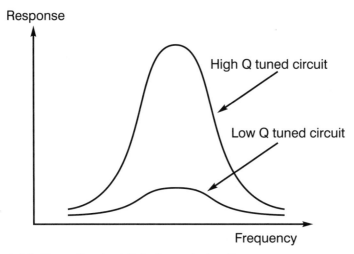

Figure 3.12 *The effect of Q in tuned circuit*

Typically an inductor may have a Q of a hundred or more, but this is very dependent upon the inductor itself. One with many turns of thin wire is likely to have a much greater loss than that of a small air-cored one using thick wire.

Low pass and high pass filters

Many circuits using inductors, capacitors and resistors for filtering applications do not just use them in the conventional series or parallel configurations. By using various circuits it is possible to construct circuits which accept signals below a certain frequency – a low pass filter – or a circuit which accepts signals above a certain frequency – a high pass filter. The very simplest circuits can be made using just a capacitor and resistor, as shown in Fig 3.13.

These circuits can be particularly useful in a number of applications, for example in filtering the output of a transmitter to ensure that harmonics above the frequency of operation are not passed to the antenna.

There are a number of important points about the curves of low and high pass filters. The first is the cut-off frequency. This is the point at which the response falls by 3 dB. Another is the rate at which the response falls when the signal is outside the pass-band. The CR circuits are not as effective as other more complicated circuits. The final rate of attenuation is only 6 dB per octave. To improve the performance the resistor can be replaced by an inductor. In this case the final rate of attenuation is 12 dB per octave. Further

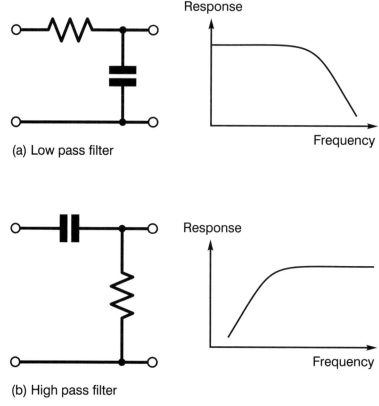

(a) Low pass filter

(b) High pass filter

Figure 3.13 *Simple CR high and low pass filters*

improvements can be made by adding further elements as shown in Figure 3.14.

Apart from the final rate of attenuation the impedance must be considered. This is because these filters have a characteristic impedance. They must be driven by the correct source impedance and given the correct load if they are to operate correctly. For many RF applications 50 ohms is a standard value. In-band ripple is also important, so is the rate at which the response starts to fall.

Not all filters have exactly the same type of response. It is possible to optimize the design so that there is minimum in-band ripple. This type of filter is called a Butterworth filter. For the fastest roll-off an elliptical filter should be used, although this type suffers from relatively large amounts of stop-band ripple. A Chebychev filter has a certain amount of in-band ripple and while it does not give as fast a roll-off as the elliptical filter it does not have the theoretical stop-band ripple, although in practice there is some. A

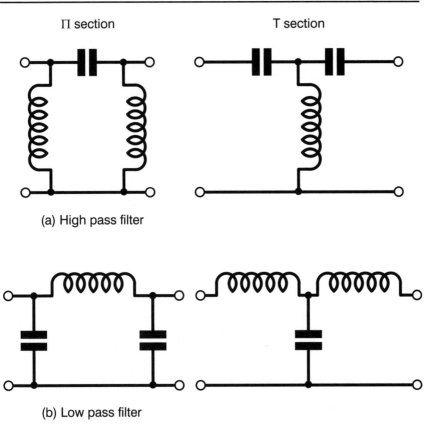

Π section   T section

(a) High pass filter

(b) Low pass filter

Figure 3.14 *High and low pass filter configurations*

Bessel filter is employed where phase considerations are important. It has a linear phase relationship, but gives the slowest roll-off.

The calculations for the component values for these filters can become complicated. The normal approach which is adopted is to refer to tables of precalculated values. The tables consist of values for different types of filter normalized to an impedance of 1 and a cut-off frequency of 1 Hz. The values can then be scaled to the required impedance and frequency to give the required performance.

Band-pass filters

Band-pass filters are widely used in receivers and other applications to allow a band of frequencies through. In a receiver the main filter is likely to be required to pass signals over a given band of frequencies, and reject others outside this. Typically a filter for receiving AM transmissions may be 6 kHz wide. In another application it may be necessary to have a filter which is possibly 500 kHz wide. Although the applications may be very different the same basic principles apply.

A very basic form of band-pass filter is a simple tuned circuit, but this is not always ideal because it is unlikely to possess the correct characteristics. In most instances the filter will be required to have a flat response over a given bandwidth, falling away rapidly outside the limits.

When designing any circuit which uses a filter it is necessary to specify what performance is required. An ideal filter may have a response like that shown in Figure 3.15, but in reality it is not possible to achieve this.

The response of a real filter like one used in a receiver to give it its selectivity is different and is shown in Figure 3.16. From this it can be seen that the response curve does not change instantly from accepting signals to rejecting them. Also it does not have an infinite rejection of unwanted signals. Nevertheless many filters achieve very good results.

The first point which needs to be noted on any filter is its pass-band. This is the bandwidth over which the filter accepts signals. The limits of the pass-band are normally taken to be the points where the response has fallen by 6 dB. For a filter used in a communications receiver to receive an AM signal on the short wave bands the bandwidth may be 6 kHz.

The stop-band is also of interest. For this the bandwidth for a given attenuation is taken. The normal value for this is 60 dB although again other values may be taken. As a result the bandwidth and attenuation levels should be specified. Typically a filter might be specified as having a stop-band bandwidth of 6 kHz at −60 dB.

The rate at which the filter reaches its final attenuation is of interest. Even though it may have the correct pass-band, if it does not reach its final

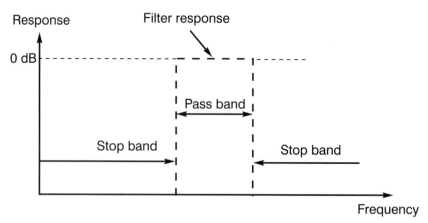

Figure 3.15 *An ideal filter response*

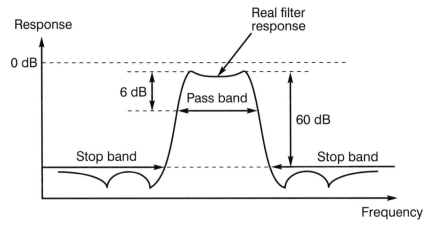

Figure 3.16 *A typical filter response*

attenuation fast enough, it will let interference through. The shape factor is a measure of the rate at which it reaches its stop-band. This is taken as the stop-band bandwidth divided by the pass-band bandwidth. Again the levels of attenuation must be given. Taking the figures already used as an example the shape factor would be 2.2:1 at 6/60 dB.

It is also noticed that the filter response is not flat. From Figure 3.16 it can be seen that there is ripple within the pass-band, and also in the stop-band. Pass-band ripple is generally only a few decibels and this is often quoted in the specification for the filter. Typical values for filter may be one or two decibels. The stop-band ripple is also of importance and can easily be tens of decibels.

Filters also introduce a certain amount of loss. This can be important for circuit designers because the effect of any losses has to be taken into consideration when calculating signal levels at certain points in the circuit. It may be in some circumstances that there will be a balance between the degree of rejection of unwanted signals and the loss of the filter. Generally the greater the rejection of unwanted signals, the more stages are required in the filter, and the greater the in-band loss.

The final rejection of the unwanted signals is another important parameter. Any filter will only be able to reject signals outside the pass-band by a certain amount. Normally the figure for the final rejection is taken so that any ripple in the stop-band does not bring the response above the figure for the final rejection. When designing and constructing a filter it is necessary to ensure that there are not any leakage paths between the input and the output of the filter. Sometimes there may be a small amount of capacitance between the two connections and this can allow signals to leak across the filter, thereby reducing the level of the final rejection.

LC filters

LC filters are found in most receivers. They consist of inductors and capacitors and are often in the form of tuned transformers used to couple the individual stages together, as shown in Figure 3.10. In many portable sets used for broadcast reception two or three of these transformers are used to give the required degree of selectivity, although for a good quality communications receiver they are not sufficient on their own.

A further disadvantage of LC filters is that they require alignment. If transformers are used then each one has its own ferrite tuning core. This is required so that the resonant frequency of the circuit can be tuned to exactly what is required. These have to be tuned to the correct frequency during the

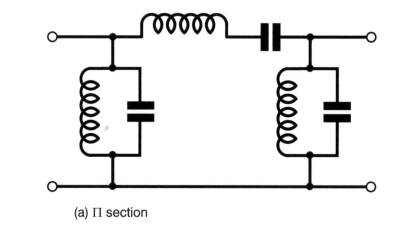

(a) Π section

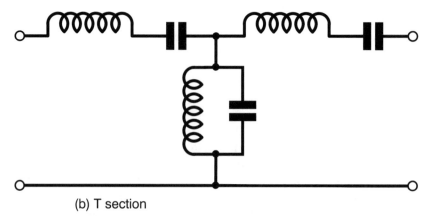

(b) T section

Figure 3.17 *LC band-pass filters*

production of the set. Normally each tuned circuit is slightly offset from the others so that the correct bandwidth is achieved. If they were all placed on exactly the same frequency then the bandwidth would be too narrow.

Other forms of LC filter can be designed, with a similar topology to the high or low pass filters, using a form of ladder as shown in Figure 3.17. These filters tend not to be used as widely in receiver applications because their design and alignment is more difficult than the inter-stage tuned transformers.

Quartz crystals

Quartz crystals play in important role in radio technology. They are widely used as very high Q resonant elements for use in filters and they are ideal in highly stable oscillators. They are able to give degrees of performance which cannot be achieved using any other component of a similar price.

Quartz uses the piezo-electric effect for its operation. This effect converts a mechanical stress in the crystal into an electric potential, and vice versa. This links the mechanical resonances of the crystal into the electrical circuit. In other words the piezo-electric effect is used to convert electrical impulses into a mechanical stress which is subject to the mechanical resonances and then these vibrations are linked back into the electrical circuit. In this way crystals are able to provide an exceptional performance. Qs of several tens of thousand are typical, a great improvement over LC circuits which only give a few hundred at the very best.

The individual components are manufactured from large man-made crystals. These are several inches long and about two inches in diameter and have a hexagonal cross-section. The individual crystal resonators are cut from the large crystal. The angle at which the cuts are made determines many of the major characteristics of the crystal. The cut most commonly used for radio frequency applications is called the AT cut. This offers a very good compromise between reducing the unwanted responses and keeping other properties like temperature stability within suitable limits.

Individual crystal blanks are cut from the main crystal using diamond wheels in view of the hardness of quartz. The blanks are in the form of discs about the size of a small thin coin, although there is a large amount of variation because actual size governs the frequency of operation. Once the blanks have been obtained they are lapped to approximately the correct thickness. The lapping paste is usually made from silicon carbide or aluminium oxide and it is very fine so that an extremely good finish can be obtained. Despite this the blank still contains discontinuities and a final stage of chemical etching is performed to give a near perfect finish.

Once complete, the next stage is to mount the crystal. Silver or gold contacts are chemically deposited onto either side of the blanks. The amount of metal which is deposited onto the blanks acts as a final trim to ensure exactly the right frequency is obtained. The final stage in the manufacture is

Figure 3.18 *A selection of quartz crystals*

(a) Longitudinal mode

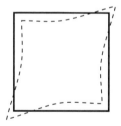

For the sake of clarity, the amplitude
of the vibration is greatly exaggerated

(b) Low frequency face-shear mode

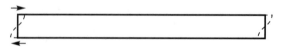

(c) High frequency shear mode

Figure 3.19 *Vibration modes of a crystal*

to mount the crystal into a can which is evacuated or filled with an inert gas to ensure that effects of ageing are minimized.

Any crystal has a number of different resonances, dependent upon the way in which it vibrates. Figure 3.19 shows some of the modes of vibration which can be excited. It is obviously important that only one mode of vibration can be excited, otherwise spurious responses will be seen. Fortunately the way in which the blank is cut from the original crystal can reduce this. The angle of the faces relative to the original crystal axes determines many of the properties. By choosing the correct cut, it is possible to reduce or eliminate most of the unwanted responses. Even so it is a wise precaution in any oscillator circuit to add an LC tuned circuit to ensure that only the correct response is excited.

It is often useful to be able to analyse the way in which a crystal responds by generating an equivalent circuit. This simulates the crystal in terms of other components so that its electrical performance can be predicted. The equivalent circuit of a crystal is shown in Figure 3.20. In this C_1 represents

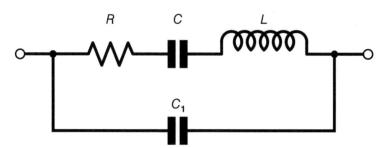

Figure 3.20 *Equivalent circuit of a crystal*

the capacitance between the electrodes. L, C, and R represent the vibrational characteristics of the crystal. L results from the mass of the material, C comes from the compliance, and R results from the losses most of which are frictional.

It can be seen that there are two ways in which the crystal can resonate. One is in the series mode, where L and C form a series resonant circuit. In this mode the impedance seen by the external circuit at resonance is low, being determined by the effective resistance in the circuit. In this mode the external circuit has little effect on the crystal.

The crystal also resonates in a parallel mode. This occurs when the combination of L, C and R have an inductive reactance which equals C_1 plus the capacitance of the external circuit. For this reason crystals which are to operate in the parallel resonant mode have a load capacitance specified. The

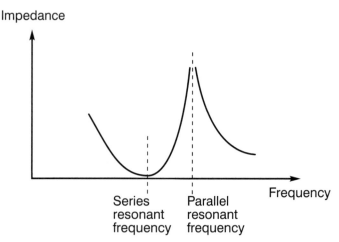

Figure 3.21 *Impedance characteristic of a crystal*

value of 30 pF is commonly used, although other values including 20 pF are popular, especially for higher frequencies.

Crystals can be operated in either the parallel or series resonant mode as seen by the impedance characteristic shown in Figure 3.21. The difference between the two frequencies is quite small, often around 1%. Normally the parallel resonant mode is used, but on occasions the series mode may also be used. The mode used depends upon the circuit and the position of the crystal within it. In the case of oscillators different circuits are used for different modes.

Crystals are generally manufactured for operation below about 30 MHz. Above this the crystal elements become very small and fragile. As a result they are not always easy to manufacture, and this increases costs considerably. When it is necessary for them to operate at high frequencies, they can be made to operate in an overtone mode. Normally the third or fifth overtones (i.e. three or five times the fundamental frequency). When crystals operate in an overtone mode, they do not vibrate at their fundamental frequency. Instead they adopt a new mode of operation, vibrating at virtually a multiple of the fundamental. It should be noted that the overtone frequency is not quite the same as a multiple of the fundamental. To ensure the correct mode of operation is selected it is necessary to ensure that the oscillator circuit is tuned so that feedback only occurs at the right frequency.

Crystal filters

In view of the very high levels of Q which quartz crystals possess they are ideal for use in filters where they are able to give far higher levels of selectivity than those which can be obtained with LC filters. Being more expensive than LC filters they are not normally used in broadcast sets. Instead they are often found in professional and amateur communications receivers and transceivers where high levels of performance are required.

A single crystal on its own does not give an ideal response curve. The response is not symmetrical, and has a very narrow peak. This is not ideal for all types of transmission as they need a certain bandwidth to accept all of the signal. To obtain a much better performance in terms of the ultimate rejection and required bandwidth, a number of crystals are used together. One popular circuit is called a half lattice filter, and this is shown in Figure 3.22. The

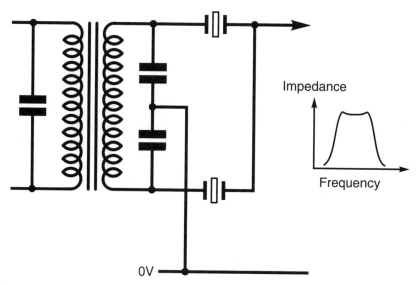

Figure 3.22 *A half lattice crystal filter circuit*

response curve for this is far more even and is shown in the diagram, although there is some ripple. To achieve the wider bandwidth the resonant frequencies of the two crystals are spaced apart slightly. As a rough guide the bandwidth of the filter at the −3 dB points is approximately 1.5 times the difference in the resonant frequencies of the crystals.

Two important factors of any filter are the rate at which the response falls away, and the ultimate rejection. These can both be improved by adding further stages to the filter. A two pole filter, i.e. one with two crystals like that shown in the diagram, is rarely sufficient to meet the requirements in a receiver. Accordingly six or eight pole versions are more common.

Monolithic crystal filters

Monolithic crystal filters are the crystal equivalent of an integrated circuit. Instead of having separate crystals on a circuit board, it is possible to place a complete filter on a single crystal. In essence the filter is made up by placing two sets of electrodes on opposite sides of a single AT cut crystal. The coupling between these electrodes acts in such a way that a highly selective filter is made.

Figure 3.23 shows a basic monolithic filter. When a signal is placed across one set of electrodes this is converted into mechanical vibrations using the piezo-electric effect. These vibrations travel across the crystal and are affected by its mechanical resonances. The vibrations then appear as the filtered signal at the other two electrodes.

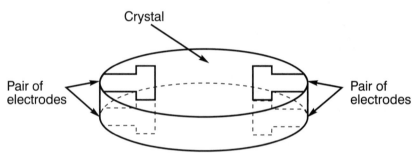

Figure 3.23 *A monolithic crystal filter*

The operation of the filter can be seen a little more easily by looking at the equivalent circuit shown in Figure 3.24. Here L_1 and C_1 determine the resonant frequency of the filter. As there are two sections to the filter both must be matched, although in practice both sides of the filter will have a very slightly different resonant frequency to give the required bandwidth. Other components are also present. L_2 is the internal coupling between the two resonant circuits. C_2 is the input and output capacitance which arises from the capacitance between the electrodes on the crystal. Finally C_3 is a parasitic capacitance across the filter from the input to the output. This must be kept to the absolute minimum, otherwise signals appearing at the input will be able to bypass the filter.

Monolithic crystal filters are generally manufactured for operation below about 30 MHz. Above this the crystals become too small to manufacture successfully. When filters are required for operation above this they are often designed to operate in an overtone mode.

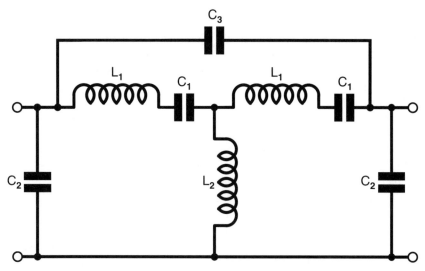

Figure 3.24 *Equivalent circuit of a monolithic crystal filter*

Mechanical filters

Crystal filters are by no means the only type of high performance filter which can be found in receivers. Mechanical filters are another option which found widespread use a number of years ago. Now that the performance of crystal filters has improved and the relative cost has fallen, mechanical filters are seldom used in new designs these days.

The mechanical filter uses the same basic principles as those used for crystal filters. Electrical vibrations are converted into mechanical ones which use mechanical resonances to achieve high degrees of selectivity.

The mechanical filter uses individual components to perform the various functions required in the filter. At either end of the filter assembly special transducers convert the electrical signals into mechanical vibrations. These signals are applied to a series of discs which are made to be mechanically resonant at the required frequency. Individual discs are linked by coupling rods. In the same way that a number of crystals are used to improve the overall selectivity and give the required bandwidth, the same principle is used for these mechanical filters. Once the mechanical vibrations have passed along the filter, they are converted back to electrical signals by a second transducer.

Mechanical filters are limited in their frequency of operation to about 500 kHz, and the resonant frequency drifts with temperature. However, one of their main advantages is that exceedingly narrow bandwidths can be achieved with relative ease.

Ceramic filters

A variety of different substances, for example certain ceramics, exhibit the piezo-electric effect, although it is not combined with such a sharp mechanical resonance. Even so they give a much better performance than their LC counterparts.

Ceramic filters operate in exactly the same way as quartz filters, the signal being linked to the mechanical resonances in the substance. As the resonance is less sharp, the bandwidths of these filters is wider and the shape factor is poorer.

Filters made from these ceramics are much cheaper to manufacture than quartz ones, and as a result they find widespread use in many areas. Many millions are used each year in broadcast receivers where they are ideal in terms of cost and performance. They are never used in performance sets as the prime source of filtering, although they are sometimes included where a small degree of filtering is used. However, as the performance of these filters is being improved all the time, their use is likely to increase.

Frequency standards

Although crystals are widely used in filters, they are also widely used in oscillators enabling very high degrees of stability to be obtained. In the oscillator circuit the crystal replaces the ordinary resonant element, as shown in Figure 3.25.

The accuracy and stability of a crystal oscillator is very much better than an LC tuned oscillator. Typically a crystal oscillator may have an accuracy of a few parts in 10^6. This is quite adequate for many applications including reference oscillators in radio receivers and many other items. However, in some instances greater accuracy is required, for example in frequency counters, signal generators and many other applications. In cases like these the stability can be greatly improved by keeping the oscillator in a temperature-controlled environment or oven. These ovens are generally a cube often about $3\,cm^3$. They contain the oscillator circuitry in a thermally insulating package. The temperature is maintained by a small thermo-statically controlled heater. These ovens are capable of giving accurate signals to within a few parts in 10^7.

If even greater accuracy is required there are a number of other options. Rubidium standards are particularly accurate (and expensive) and are sometimes used in laboratories. Another option is to use an off-air standard. There are a number of standard frequency transmissions around the world that are maintained to a very high degree of accuracy. Some are on the short wave bands usually on 2.5, 5.0, 10.0 and 15.0 MHz, but those of interest for generating laboratory standard signals are generally found on the VLF portion of the spectrum around 50 kHz. Two examples of these are MSF at

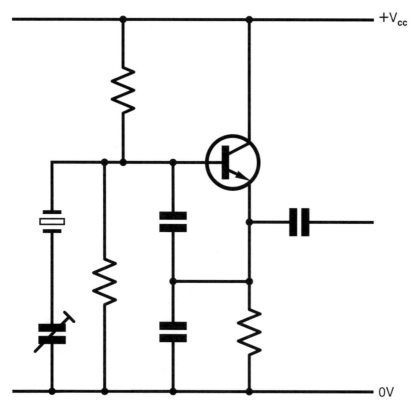

Figure 3.25 *A crystal oscillator*

Rugby in the UK on 60 kHz and DCF77 in Germany on 77.5 kHz. The signals from these stations can be used to lock an oscillator onto the correct frequency, thereby generating a very accurate reference. The main drawback of these off-air standards is that changes in the ionospheric conditions mean that there are small short-term frequency variations. Even so they are still capable of providing a very accurate frequency source, and are particularly good over a long period of time.

4 Modulation

Radio signals are used to carry information. Some are used for carrying sound broadcasting, others are used for mobile telephone communications. In fact there is an almost infinite variety of uses for radio signals, but in all cases they are used to carry information in one form or another.

To be able to transmit any form of information a radio signal or carrier is first generated. The information in the form of audio, data, or other form of signal is used to modify or modulate the carrier and in this way the information is superimposed onto the carrier and is transmitted to the receiver. Here the information is removed from the radio signal and reconstituted in its original format by a process known as demodulation.

There are many ways in which a carrier can be modulated. Each one has its own advantage and can perform well under given circumstances. Some of the simpler forms of transmission have the advantage that receivers needed to resolve them properly are not as complicated. On the other hand, other modes which need more complicated circuitry to resolve them properly may perform better in one aspect or another.

Radio carrier

The basis of any radio signal or transmission is the carrier. This consists of an alternating waveform like that shown in Figure 4.1. This is generated in the transmitter and is usually passed to the output of the transmitter where it is radiated for reception. This may be anywhere within the radio spectrum described in Chapter 2.

The carrier in itself does not convey any sound. It is purely the radio signal which carries the information or modulation which is used to modify it. There are many ways of achieving this. Some are very involved, requiring complicated circuitry to encode and decode them. Others are very much simpler.

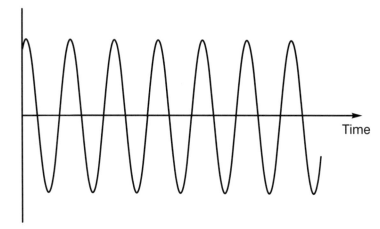

Figure 4.1 *An alternating waveform*

Morse

Morse is the oldest and simplest way of transmitting information using radio. Yet despite its age it still has several advantages over other forms of transmission, and is therefore still in fairly widespread use today.

One of its advantages is its simplicity. It only consists of a carrier wave which is turned on and off as shown in Figure 4.2. The characteristic dots and

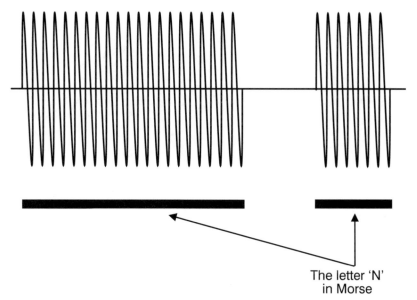

The letter 'N'
in Morse

Figure 4.2 *A Morse signal*

Table 4.1   The Morse code

A	• –	N	– •
B	– • • •	O	– – –
C	– • – •	P	• – – •
D	– • •	Q	– – • –
E	•	R	• – •
F	• • – •	S	• • •
G	– – •	T	–
H	• • • •	U	• • –
I	• •	V	• • • –
J	• – – –	W	• – –
K	– • –	X	– • • –
L	• – • •	Y	– • – –
M	– –	Z	– – • •
1	• – – – –	6	– • • • •
2	• • – – –	7	– – • • •
3	• • • – –	8	– – – • •
4	• • • • –	9	– – – – •
5	• • • • •	0	– – – – –

Punctuation

Full Stop	• – • – • –
Comma	– – • • – –
Question Mark (?)	• • – – • •
Equals sign (=)	– • • • –
Stroke (/)	– • • – •
Mistake	• • • • • • • •

Procedural characters
Procedural characters made up of two letters are sent as a single letter
with no break between them.

Start of Work (CT)	– • – • –
Invitation to Transmit (KN)	– • – – •
End of Work (VA)	• • • – • –
End of Message (AR)	• – • – •
Invitation to Transmit (K)	– • –
Invitation to a Particular Station to Transmit (KN)	– • – – •

Timing

length of a dash	3 dots
space between dots and dashes	1 dot
space between letters	3 dots
space between words	5 dots

dashes are defined by the length of time the transmission is left on. The dots and dashes then make up the required letters according to the Morse code given in Table 4.1. From this it can be seen that the signal transmitted in Figure 4.1 is a dash and a dot which makes the letter 'A'.

The code has been used in this form for over a century, but the first code devised by Samuel Morse had a number of shortcomings. Accordingly he revised it, and it has been used in this form ever since. As it is used worldwide, it is often called the International Morse Code. This distinguishes it from other codes which are needed for countries like Russia which have a different alphabet.

The simplicity of the Morse code and its implementation means that equipment for sending it can be more basic than many other modes. This can be an advantage for amateur radio enthusiasts as it enables them to build their equipment more easily.

Morse has a number of technical advantages. Its relatively slow signalling rate means that it occupies a narrow bandwidth. As a result very narrow filters can be used to cut out most of the interference. It is also found that because the human brain only has to detect the presence or absence of a signal, it can be read at a lower level than a signal carrying speech. These two factors together mean that Morse can be copied at much lower signal levels than other forms of transmission.

If Morse is received by an ordinary domestic portable radio receiver it is simply heard as a series of clicks and pops as the signal is turned on and off. To make the characteristic tone, a circuit called a beat frequency oscillator (BFO) is needed. This circuit generates a signal which mixes or beats with the incoming signal to make the characteristic Morse sounds of a Morse signal.

Amplitude modulation

Morse code has its advantages but we have all become accustomed to hearing music and speech over the radio. There are a number of ways in which a carrier can be modulated to take an audio signal. The most obvious way is to change its amplitude in line with the variations in intensity of the sound wave. In this way the overall amplitude or envelope of the carrier is modulated to carry the audio signal as shown in Figure 4.3. Here the envelope of the carrier can be seen to change in line with the modulating signal.

Amplitude modulation or AM is one of the most straightforward methods of modulating a signal. Demodulation, or the process where the radio frequency signal is converted into an audio frequency signal, is also very simple. It only requires a simple diode detector circuit like that shown in Figure 4.4. In this circuit the diode rectifies the signal, only allowing one half of the alternating radio frequency waveform through. A capacitor is used to

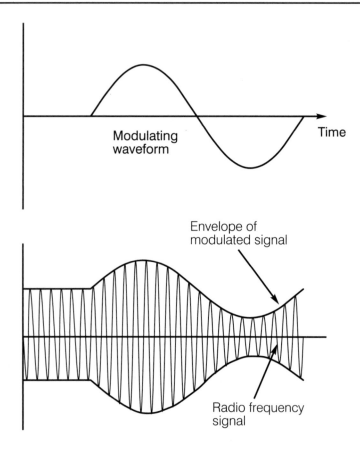

Figure 4.3 *An amplitude modulated signal*

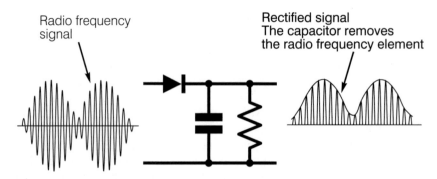

Figure 4.4 *A simple diode detector circuit*

remove the radio frequency parts of the signal, leaving the audio waveform. This can be fed into an amplifier after which it can be used to drive a loudspeaker. As the circuit used for demodulating AM is very cheap, it enables the cost of receivers for AM to be kept low.

AM has the advantage of simplicity, but it is not the most efficient mode to use, both in terms of the amount of spectrum it takes up and the usage of power. It is for this reason that it is rarely used for communications purposes, although it is used for VHF aircraft communications. However, it is still widely used on the long, medium, and short wave bands for broadcasting because of its simplicity.

To find out why it is inefficient it is necessary to look at a little theory behind the operation of AM. When a radio frequency signal is modulated by an audio signal the envelope will vary. The level of modulation can be increased to a level where the envelope falls to zero and then rises to twice the unmodulated level. Any increase on this will cause distortion. As this is the maximum amount of modulation possible it is called 100% modulation.

Even with 100% modulation the utilization of power is very poor. When the carrier is modulated sidebands appear at either side of the carrier, and it is these signals which contain the information about the audio modulation. To look at how the signal is made up and the relative powers, take the simplified case where the 1 kHz tone is modulating the carrier. In this case two signals will be found 1 kHz either side of the main carrier, as shown in Figure 4.6. When the carrier is fully modulated, i.e. 100%, the power in each sideband

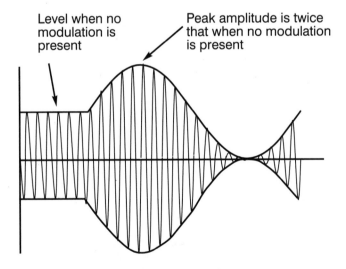

Figure 4.5 *Fully modulated signal*

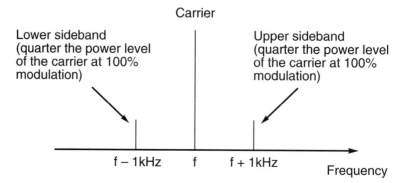

Figure 4.6 *Spectrum of a signal modulated with a 1 kHz tone*

rises to half the voltage level of the main carrier. In terms of power this means that each sideband is just a quarter of the level. During the modulation process the carrier remains constant and is only needed as a reference during the demodulation process.

Not only is AM wasteful in terms of power, it is also not very efficient in its use of spectrum. If the 1 kHz tone is replaced by a typical audio signal

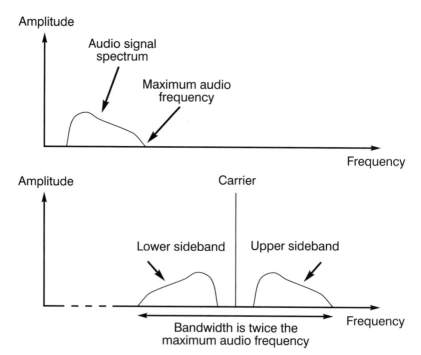

Figure 4.7 *Spectrum of a signal modulated with speech or music*

made up of a variety of sounds with different frequencies then each frequency will be present in each sideband. Accordingly the sidebands spread out either side of the carrier as shown and the total bandwidth used is equal to twice the top frequency which is transmitted. In the crowded conditions found on many of the short wave bands today, this is a waste of space, and other modes of transmission which take up less space are often used.

Modulation index

It is often necessary to define the level of modulation which is applied to a signal. A factor or index known as the modulation index is used for this. When expressed as a percentage it is the same as the depth of modulation. In other words it can be expressed as:

$$M = \frac{\text{RMS value of modulating signal}}{\text{RMS value of unmodulated signal}}$$

The value of the modulation index must not be allowed to exceed one (i.e. 100% in terms of the depth of modulation) otherwise the envelope becomes distorted and the signal will 'splatter' either side of the wanted channel, causing interference and annoyance to other users.

Single sideband

One of the modes widely used for communications traffic is called single sideband (SSB). This is a derivative of AM, but by manipulating the signal in the transmitter, the disadvantages of AM can be removed to give a highly efficient mode of transmission.

There are two main stages in the generation of a single sideband signal. The first is that the carrier is removed. This does not contribute to carrying the sound information, and is only used during the demodulation process. As a result it is possible to remove it in the transmitter as shown in Figure 4.8, enabling power to be saved.

It is also found that only one sideband is needed. Both the upper and the lower sidebands are exact mirror images of one another, and either can be used equally well for conveying the sound information. If two sidebands are present on a signal twice the bandwidth is used. To save on the use of bandwidth it is possible to remove one sideband without degrading the signal. This can be done because both sidebands carry exactly the same information; they are exact mirror images of one another. The other advantage of transmitting only one sideband is that filter bandwidths can be reduced in the receiver to cut out more unwanted interference and give better reception.

To demodulate the signal the carrier has to be reintroduced in the receiver using a beat frequency oscillator (BFO), which may also be called

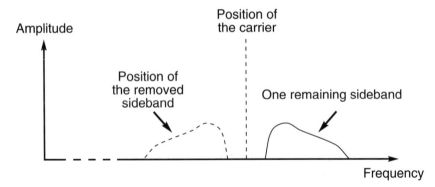

Figure 4.8 *Spectrum of a single sideband signal*

a carrier insertion oscillator (CIO). The BFO must be on the correct frequency relative to the sideband being received. Any deviation from this will cause the pitch of the recovered audio to change. While errors of up to about 100 Hz are acceptable for communications applications including amateur radio, if music is to be transmitted the carrier must be reintroduced on exactly the correct frequency. This can be accomplished by transmitting a small amount of carrier, and using circuitry in the receiver to lock onto this.

As either sideband can be used equally well, a convention is needed for which sideband to use. In this way receivers can be set up to expect the received sideband. If this is not done then the receiver has to be continually switched between upper and lower sidebands. Commercial operators use the upper sideband in all instances, whereas radio amateurs use the upper sideband on bands above 10 MHz and lower sideband below this. Although it is perfectly possible to use SSB for frequencies above 30 MHz it is infrequently used on these frequencies. Signal propagation and the way the bands are used mean that other modes are often more suitable.

It is often necessary to define the output power of a single sideband transmitter. As the output is continually varying and dependent upon the level of modulation at any instant, a measure known as the peak envelope power (PEP) is used. This is the peak level of power of the transmitted signal and includes the sideband plus any pilot carrier which may be included.

Frequency modulation

The most obvious method of applying modulation to a signal is to superimpose the audio signal onto the amplitude of the carrier. However, this is by no means the only method which can be employed. It is also possible

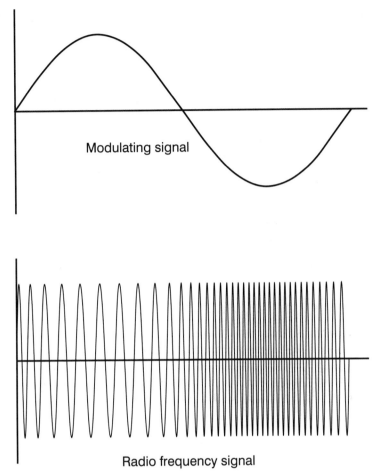

Modulating signal

Radio frequency signal

Figure 4.9 *A frequency modulated signal*

to vary the frequency of the signal to give frequency modulation or FM. It can be seen from Figure 4.9 that the frequency of the signal varies as the voltage of the modulating signal changes.

The amount by which the signal frequency varies is very important. This is known as the deviation and is normally quoted as the number of kilohertz deviation. As an example the signal may have a deviation of ±3 kHz. In this case the carrier is made to move up and down by 3 kHz.

Broadcast stations in the VHF portion of the frequency spectrum between 88.5 and 108 MHz use large values of deviation, typically ±75 kHz. This is known as wide-band FM (WBFM). These signals are capable of supporting high quality transmissions, but occupy a large amount of bandwidth. Usually

200 kHz is allowed for each wide-band FM transmission. For communications purposes less bandwidth is used. Narrow-band FM (NBFM) often uses deviation figures of around ±3 kHz or slightly more.

FM is used for a number of reasons. One particular advantage is its resilience to signal level variations. The modulation is carried only as variations in frequency. This means that any signal level variations will not affect the audio output, provided that the signal does not fall to a level where the receiver cannot cope. As a result this makes FM ideal for mobile or portable applications where signal levels are likely to vary considerably. The other advantage of FM is its resilience to noise and interference. It is for this reason that FM is used for high quality broadcast transmissions.

To demodulate an FM signal it is necessary to convert the frequency variations into voltage variations. This is slightly more complicated than demodulating AM, but it is still relatively simple to achieve. Rather than just detecting the amplitude level using a diode, a tuned circuit has to be incorporated so that a different output voltage level is given as the signal changes its frequency.

Modulation index and deviation ratio

In just the same way that it is useful to know the modulation index of an amplitude modulated signal, the same is true for a frequency modulated signal. The modulation index is equal to the ratio of the frequency deviation to the modulating frequency. The modulation index will vary according to the frequency which is modulating the transmitted carrier and the amount of deviation. However, when designing a system it is important to know the maximum permissible values. This is given by the deviation ratio and is obtained by inserting the maximum values into the formula for the modulation index.

$$D = \frac{\text{Max. deviation frequency}}{\text{Max modulation frequency}}$$

For a VHF FM sound broadcast transmitter the maximum deviation is 75 kHz and the maximum modulation frequency is 15 kHz giving a deviation ratio of 5.

Sidebands

Any signal which is modulated produces sidebands. In the case of an amplitude modulated signal they are easy to determine, but for frequency modulation the situation is not quite as straightforward. For small values of modulation index, when using narrow-band FM, and FM signal consists of the carrier and the two sidebands spaced at the modulation frequency either side of the carrier. This looks to be the same as an AM signal, but the difference is that the lower sideband is out of phase by 180°.

As the modulation index increases it is found that other sidebands at twice the modulation frequency start to appear. As the index is increased further other sidebands can also be seen. It is also found that the relative levels of these sidebands change, some rising in level and others falling as the modulation index varies. This makes prediction of the exact levels of all the sidebands more difficult than for AM. If a mathematical calculation of the levels of the sidebands is required, a function known as a Bessel function or series needs to be calculated.

Bandwidth

In the case of an amplitude modulated signal the bandwidth required is twice the maximum frequency of the modulation. While the same is true for a narrow-band FM signal, the situation is not true for a wide-band FM signal. Here the required bandwidth can be very much larger, with detectable sidebands spreading out over large amounts of the frequency spectrum. Usually it is necessary to limit the bandwidth of a signal so that it does not unduly interfere with stations either side.

While it is possible to limit the bandwidth of an FM signal, this should not introduce any undue distortion. To achieve this it is normally necessary to allow a bandwidth equal to twice the maximum frequency of deviation plus the maximum modulation frequency. In other words for a VHF FM broadcast station this must be $(2 \times 75) + 15\,\text{kHz}$, i.e. $180\,\text{kHz}$. In view of this a total of $200\,\text{kHz}$ is usually allowed, enabling stations to have a small guard band and their centre frequencies on integral numbers of $100\,\text{kHz}$.

Improvement in signal to noise ratio

It has already been mentioned that FM gives a better signal to noise ratio than AM. Most noise appears as amplitude noise and this can be removed by limiting the signal. It can also be shown that FM is inherently less susceptible to noise. The greater the deviation the better the noise performance. In fact when comparing an AM signal to an FM one an improvement equal to $3D^2$ is obtained where D is the deviation ratio.

Pre-emphasis and de-emphasis

An additional improvement in signal to noise ratio can be achieved if the audio signal is pre-emphasized. To achieve this the lower level high frequency sounds are amplified to a greater degree than the lower frequency sounds before they are transmitted. Once at the receiver the signals are passed through a network with the opposite effect to restore a flat frequency response.

To achieve the pre-emphasis the signal is passed through a capacitor resistor (CR) network. At frequencies above the cut-off frequency the signal increases in level by 6 dB per octave. Similarly at the receiver the response falls by the same amount.

Both the receiver and transmitter networks must match one another. In the UK the CR time constant is chosen to be 50 μs. For this the break frequency f_1 is 3183 Hz. In North America values of 75 μs with a break frequency of 2.1 kHz are used.

Pre-emphasizing the audio for an FM signal is effective because the noise output from an FM system is proportional to the audio frequency. In order to reduce the level of this effect, the audio amplifier in the receiver must have a response which falls with frequency. In order to prevent the audio signal from losing the higher frequencies, the transmitter must increase the level of the higher frequencies to compensate. This can be achieved because the level of the high frequency sounds is usually less than those lower in frequency.

Frequency shift keying

Digital data cannot be transmitted directly over a radio link very successfully. The square edges of the data require a wide bandwidth, and the transmission path is likely to distort the signal so that large numbers of data errors are encountered. To overcome this a slightly different approach is adopted.

Many signals heard on the bands will employ a system called frequency shift keying or FSK. Here the frequency of the signal is changed from one frequency to another, one frequency counting as the digital one (mark) and the other as a digital zero (space). By changing the frequency of the signal between these two frequencies it is possible to send data over the radio.

FSK is widely used on the HF bands. To generate the audio tone required from the receiver, a beat frequency oscillator must be used. Accordingly to obtain the correct audio ones the receiver must be tuned to the correct frequency.

At frequencies in the VHF and UHF portion of the spectrum a slightly different approach is adopted. An audio tone is used to modulate the carrier and the audio is shifted between the two frequencies. Although the carrier can be amplitude modulated, frequency modulation is virtually standard. Using audio frequency shift keying (AFSK) the tuning of the receiver becomes less critical.

When the data signal leaves the receiver it is generally in the form of an audio signal switching between two tones. This needs to be converted into the two digital signal levels, and is achieved by a unit called a modem. This stands for MOdulator/DEModulator. Audio tones fed into the modem from the receiver generate the digital levels required for a computer or other equipment to convert into legible text. Conversely it is able to convert the

Table 4.2 Murray Code

Lower case	Upper case	Code Element					Decimal value
		5(MSB)	4	3	2	1	
A	-	0	0	0	1	1	3
B	?	1	1	0	0	1	25
C	'	0	1	1	1	0	14
D	$ AB	0	1	0	0	1	9
E	3	0	0	0	0	1	1
F	! %	0	1	1	0	1	13
G	& @	1	1	0	1	0	26
H	£	1	0	1	0	0	20
I	8	0	0	1	1	0	6
J	' Bell	0	1	0	1	1	11
K	(	0	1	1	1	1	15
L	)	1	0	0	1	0	18
M	.	1	1	1	0	0	28
N	,	0	1	1	0	0	12
O	9	1	1	0	0	0	24
P	0	1	0	1	1	0	22
Q	1	1	0	1	1	1	23
R	4	0	1	0	1	0	10
S	Bell !	0	0	1	0	1	5
T	5	1	0	0	0	0	16
U	7	0	0	1	1	1	7
V	; =	1	1	1	1	0	30
W	2	1	0	0	1	1	19
X	1	1	1	1	0	1	29
Z	" +	1	0	0	0	1	17
Space		0	0	1	0	0	4
CR		0	1	0	0	0	8
LF		0	0	0	1	0	2
Figure Shift		1	1	0	1	1	27
Letter Shift		1	1	1	1	1	31
Blank		0	0	0	0	0	0

AB = Answer back or WHU (Who Are You?)
Upper case characters may vary in some cases as indicated on the chart. Also upper case F, G, and H are not often used

Table 4.3   ASCII (American Standard Code for Information Interchange)

Hex	Decimal	Keypress	Result
00	00	CTRL @	NUL
01	01	CTRL A	SOH
02	02	CTRL B	STX
03	03	CTRL C	ETX
04	04	CTRL D	EOT
05	05	CTRL E	ENQ
06	06	CTRL F	ACK
07	07	CTRL G	BEL
08	08	CTRL H	BS
09	09	CTRL I	HT
0A	10	CTRL J	LF
0B	11	CTRL K	VT
0C	12	CTRL L	FF
0D	13	CTRL M	CR
0E	14	CTRL N	SO
0F	15	CTRL O	SI
10	16	CTRL P	DLE
11	17	CTRL Q	DC1 (X on)
12	18	CTRL R	DC2
13	19	CTRL S	DC3 (X off)
14	20	CTRL T	DC4
15	21	CTRL U	NAK
16	22	CTRL V	SYN
17	23	CTRL W	ETB
18	24	CTRL X	CAN
19	25	CTRL Y	EM
1A	26	CTRL Z	SUB
1B	27	CTRL [	ESC
1C	28	CTRL \	FS
1D	29	CTRL]	GS
1E	30	CTRL ^	RS
1F	31	CTRL -	US
20	32	space	SP
21	33	!	!
22	34	"	"
23	35	£	£
24	36	$	$

Table 4.3   Continued

Hex	Decimal	Keypress	Result
25	37	%	%
26	38	&	&
27	39	'	'
28	40	(	(
29	41	)	)
2A	42	*	*
2B	43	+	+
2C	44	,	,
2D	45	-	-
2E	46	.	.
2F	47	/	/
30	48	0	0
31	49	1	1
32	50	2	2
33	51	3	3
34	52	4	4
35	53	5	5
36	54	6	6
37	55	7	7
38	56	8	8
39	57	9	9
3A	58	:	:
3B	59	;	;
3C	60	<	<
3D	61	=	=
3E	62	>	>
3F	63	?	?
40	64	@	@
41	65	A	A
42	66	B	B
43	67	C	C
44	68	D	D
45	69	E	E
46	70	F	F
47	71	G	G
48	72	H	H
49	73	I	I
4A	74	J	J

Table 4.3   Continued

Hex	Decimal	Keypress	Result
4B	75	K	K
4C	76	L	L
4D	77	M	M
4E	78	N	N
4F	79	O	O
50	80	P	P
51	81	Q	Q
52	82	R	R
53	83	S	S
54	84	T	T
55	85	U	U
56	86	V	V
57	87	W	W
58	88	X	X
59	89	Y	Y
5A	90	Z	Z
5B	91	[	[
5C	92	\	\
5D	93	]	]
5E	94	^	^
5F	95	_	_
60	96	`	`
61	97	a	a
62	98	b	b
63	99	c	c
64	100	d	d
65	101	e	e
66	102	f	f
67	103	g	g
68	104	h	h
69	105	i	i
6A	106	j	j
6B	107	k	k
6C	108	l	l
6D	109	m	m
6E	110	n	n
6F	111	o	o
70	112	p	p

Table 4.3   Continued

Hex	Decimal	Keypress	Result
71	113	q	q
72	114	r	r
73	115	s	s
74	116	t	t
75	117	u	u
76	118	v	v
77	119	w	w
78	120	x	x
79	121	y	y
7A	122	z	z
7B	123	{	{
7C	124				
7D	125	}	}
7E	126	~	~
7F	127	del	del

digital signals into the audio tones required to modulate a transmitter to transmit data.

The speed of the transmission is important. For the receiver to be able to decode the signal it must be expecting data at the same rate as it is arriving. Accordingly a number of standard speeds are used. The unit used for this is the baud. One baud is equal to one bit per second.

Codes are used to enable the series of marks and spaces or ones and zeros to be represented as alphanumerics. Early teleprinter systems used a code called the Murray code, which is given in Table 4.2. Often ASCII (American Standard Code for Information Interchange) is used for the basis of some systems now. Also, as FSK systems are widely used on the HF bands where levels of interference are high, the systems use high levels of error checking and acknowledgements when a packet of data has been received. These systems enable the data to be received with very few errors even when some interference is present.

Phase modulation

Another form of modulation which is widely used is called phase modulation. As phase and frequency are inextricably linked both forms of modulation are often referred to by the common name of angle modulation.

To explain how phase modulation works it is first necessary to give an explanation of phase. A radio signal consists of an oscillating carrier in the form of a sine wave as shown in Figure 4.10. The amplitude follows this curve, moving positive and then negative, returning to the start point after one complete cycle. This can also be represented by the movement of a point around a circle, the phase at any given point being the angle between the start point and the point on the waveform as shown.

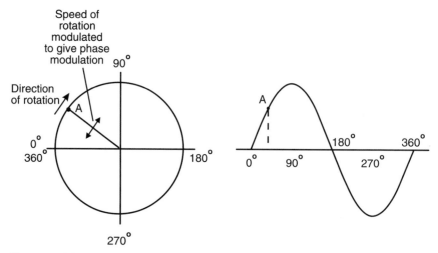

Figure 4.10 *Phase modulation*

Modulating changes the phase of the signal from what it would have been if no modulation was applied. In other words the speed of rotation around the circle is modulated about the mean value as shown in Figure 4.10. To achieve this it is necessary to change the frequency of the signal for a short time. In other words when phase modulation is applied to a signal there are frequency changes and vice versa. Phase and frequency are inseparably linked as phase is the integral of frequency. Frequency modulation can be changed to phase modulation by simply adding a CR network to the modulating signal which integrates the modulating signal.

Phase modulation is often used for multiplexed transmissions where a number of channels are applied to the same carrier. The reason for this is that with FM the channels occupying the higher section of the baseband would be subject to greater levels of noise. This fact is apparent when listening to a stereo FM broadcast. In mono, listening may be virtually noise free, whereas when the stereo capability is enabled the noise level may rise significantly if the received signal is not very strong.

Phase shift keying

Phase modulation may be used for the transmission of data. Frequency shift keying is robust, and has no ambiguities because one tone is higher than the other. However, phase shift keying (PSK) has many advantages in terms of efficient use of bandwidth, and comes into its own when used over the more stable environments available at VHF and above.

The problem with phase shift keying is that the receiver cannot know the exact phase of the transmitted signal to determine whether it is in a mark or space condition. This would not be possible even if the transmitter and receiver clocks were accurately linked because the path length would determine the exact phase of the received signal. To overcome this problem PSK systems use a differential method for encoding the data onto the carrier. This is accomplished, for example, by making a change in phase equal to a one, and no phase change equal to a zero. Further improvements can be made upon this basic system and a number of other types of phase shift keying have been developed. One simple improvement can be made by making a change in phase by $90°$ in one direction for a one, and $90°$ the other way for a zero. This retains the $180°$ phase reversal between one and zero states, but gives a distinct change for a zero. In a basic system not using this process it may be possible to lose synchronization if a long series of zeros are sent. This is because the phase will not change state for this occurrence.

There are many variations on the basic idea of phase shift keying. Each one has its own advantages and disadvantages, enabling system designers to choose the one most applicable for any given circumstances.

Pulse modulation

While the amplitude and angle modulation, together with their many derivatives, form the vast majority of transmissions, another type of modulation, known as pulse modulation, is used in many instances. The advantage of this type of modulation is that a number of different signals can be transmitted together using a system known as time division multiplexing.

Using this type of modulation, short periodic samples are transmitted. The modulating waveform is sampled as shown in Figure 4.11(a) and this information is used to change the characteristics of the pulses in one of a variety of ways. Once at the receiver the samples are used to reconstitute the original signal which can be used as required. Surprisingly few samples are needed and it can be shown that if an audio signal is sampled at a frequency just over twice the highest frequency present, then the samples will contain all the necessary information to recreate the signal in its original format. Take as an example the waveform shown in Figure 4.11(b). If the waveform was sampled once a cycle then each sample would have the same value and there

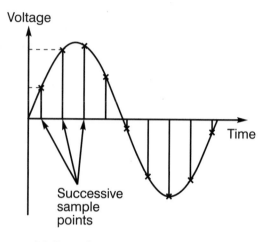

(a) Samples taken periodically

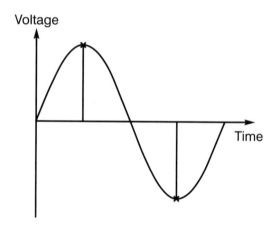

(b) Minimum number of samples

Figure 4.11 *Sampling a waveform*

would be no indication about the actual waveform. However, if the waveform is sampled at twice the frequency of the waveform, the samples alternate either side of the zero axis, and it can be deduced that it is varying and has a certain frequency.

There are a number of ways in which pulses derived from samples of a waveform can be used to modulate a carrier. Possibly the most obvious is pulse amplitude modulation (PAM) where the amplitude of the pulses

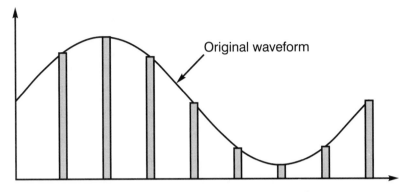

(a) Varying the pulse height to represent the waveform

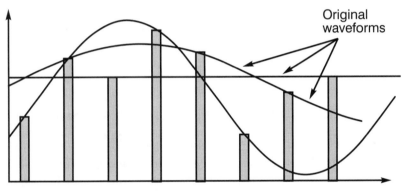

(b) Time division multiplexed pulses for three signals

Figure 4.12 *Pulse amplitude modulation*

represent the amplitude of the waveform at that point, as shown in Figure 4.12. It can also be seen that it is possible to place more than one set of modulating pulses on the signal, by spacing them apart in time. This concept is known as time division multiplexing. At the transmitting end of the link the pulses are placed onto the carrier in a given order. By taking the pulses of carrier at the receiver end in the same order, each signal can be properly reconstituted.

Another method of pulse modulation involves varying the width of the pulses according to the amplitude of the waveform. This is called pulse width modulation (PWM). An example is shown in Figure 4.13 and from this it can be seen that the width of the pulses increases in line with the instantaneous value of the waveform.

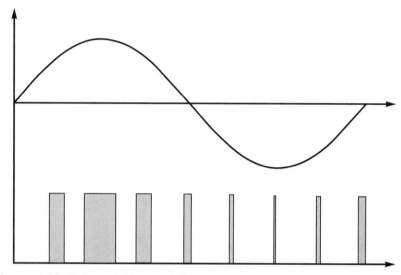

Figure 4.13 *Pulse width modulation*

Both PAM and PWM are essentially analogue methods of pulse modulating a signal. There are many advantages to digitizing the signal and using a digital train of pulses to modulate the carrier. This form of pulse code modulation (PCM) is widely used for telecommunications links. The advantage is that once the signal is digitized, the system only has to recognize whether a pulse is present or not, i.e. a logical one or a zero. If the noise level is not large then there is a very small chance of there being any errors. It is found that if the RMS noise level is 20 dB below the peak pulse level then there is a negligible chance of there being an error and at 13 dB the chance of an error becomes one in a hundred. As a result PCM systems are more resilient to noise than their analogue equivalents. This has the advantage that repeater stations to boost the signals are required less frequently giving a significant saving in costs.

To generate a PCM signal, the modulating waveform is sampled as before at distinct points. These samples are then converted into a digital format using a circuit called an analogue to digital converter. This circuit generates a binary code which in turn is converted into a group of pulses. A group of n pulses can represent 2^n different levels, i.e. if three pulses are generated then there can be 2^3 or 8 levels. In reality eight pulses are usually used for communications links giving a total of 256 levels.

To reconstitute the waveform the incoming pulses are converted into a binary code which enters a circuit called a digital to analogue converter. This regenerates the audio signal into a form of staircase waveform with steps as

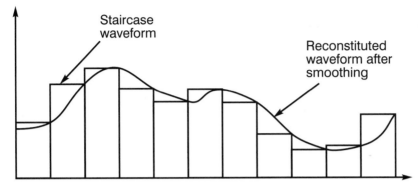

Figure 4.14 *Regenerating an analogue signal from its digital representation*

shown. The steps can be removed by adding a capacitor circuit to smooth the waveform, giving a very close approximation to the original signal.

The main disadvantage of PCM when compared to PAM or PWM is that far more pulses are transmitted, and this means that a much greater bandwidth is required. If eight pulses are used to represent each sample then the bandwidth will be eight times as much. However, the increase in noise immunity it provides usually more than outweighs the problems with the bandwidth.

Frequency hopping

In some instances, particularly in military applications, it is necessary to prevent any people apart from intended listeners from picking up a signal or from jamming it. One way of achieving this is to move the signal from one frequency to another. Frequency hopping is a well-established principle and is a form of spread spectrum transmission. In this system the signal is changed many times a second in a pseudo-random sequence from a predefined block of channels. Hop rates vary dependent upon the requirements. Typically the transmission may hop a hundred times a second, although at HF this will be much less.

The transmitter will remain on each frequency for a given amount of time before moving on to the next. There is a small dead time between the signal appearing on the next channel, during this time the transmitter output is muted. This is to enable the frequency synthesizer time to settle, and to prevent interference to other channels as the signal moves.

To receive the signal the receiver must be able to follow the hop sequence of the transmitter. This is normally achieved by 'filling' each transceiver with the required information and ensuring that an internal clock is maintained accurately. In this way it will look at the correct channel at the right time.

Frequency hopping transmissions usually use a form of digital transmission. Even when speech is used, this has to be digitized before being sent. Often it may be encrypted to further improve the security of the system. The data rate over the air has to be greater than the overall throughput. This is to allow for the dead time while the set is changing frequency.

Spread spectrum

A more versatile method of spreading the signal out over a wide band of frequencies is to use a method known as direct sequence spread spectrum. This system operates by deliberately increasing the bandwidth of the signal to spread out over a large band of frequencies where it appears as an increase in the noise level.

Like a frequency hopping signal, a direct sequence spread spectrum signal transmits the signal in the form of data which is phase modulated onto a carrier. To increase the data rate the data to be transmitted is exclusive OR'ed with pseudo-random bit sequence with a much higher data rate.

At the receiver the signal is multiplied with the pseudo-random sequence in the correct phase and this has the effect of reconstituting the signal in its non-spread form. After this it can be processed in the usual way.

The advantage of this form of spread spectrum signal is that it can be received through signals many times its own strength, and can even be received when it is well below the noise level. For covert use this gives it a significant advantage because unless the receiver has the correct pseudo-random sequence it may not even be possible to detect it is present.

Transmission codes

It is often useful to be able to define a type of transmission in an easy and abbreviated form. The system which is currently in use was devised at the World Administrative Radio Conference in 1979, and used with effect from 1982. The code consists of three characters, as shown in Table 4.4. The first is a letter and describes the type of modulation. The second is a number and defines the nature of the modulating signal. In other words the information or signal which is being carried by the transmission. The third is another letter and indicates the type of information being transmitted. For example, it may be Morse, audio information as in the case of a speech

Table 4.4   Transmission codes

First Character *Type of modulation*	Second Character *Nature of modulating signal*	Third Character *Information being transmitted*
N Emission of an unmodulated carrier	0 No modulating signal	N No information transmitted
Amplitude Modulated Carriers A Double sideband H Single sideband full carrier R Single sideband, reduced level carrier J Single sideband suppressed carrier B Independent sideband C Vestigial sideband	1 A single channel containing digital or quantized information without the use of a subcarrier 2 A single channel containing digital or quantized information with the use of a subcarrier	A Telegraphy for aural reception B Telegraphy for automatic reception C Facsimile D Data transmission E Telephony F Television W Combination of the above X Any cases not covered
Angle Modulated Carrier F Frequency modulation G Phase modulation D Combination of angle and amplitude modulation	3 A single channel containing analogue information 7 Two or more channels containing digital or quantized information 8 Two or more channels containing analogue information	
Pulse Modulation P Unmodulated sequences of pulses K Pulses modulated in amplitude L Pulses modulated in width M Pulses modulated in position or phase Q Series of pulses in which the carrier is angle modulated in the period of the pulse V Combination of the above produced by other means W A combination of two or more of the following: angle, amplitude, or pulse modulation X Any cases not covered	9 Composite signal containing digital or quantized information together with one or more channels containing analogue information 0 Any cases not covered	

transmission. As an example J3E denotes a single sideband suppressed carrier transmission, and A1A is a Morse transmission.

In some instances the bandwidth of the transmission is also included after the basic code. This consists of a four character code consisting of three numbers and a letter. A letter in the place of the decimal point denotes the units, namely H for hertz, k for kHz, M for MHz, and G for GHz. In this way a transmission occupying 12.5 kHz would be shown by 12k5.

5 Receivers

Radio receivers are very common these days, being found in a wide variety of places and in many different forms. The most obvious type of receiver is the domestic broadcast radio. Today most households have a variety of sets, ranging from portable radios up to hi-fi tuners. Car radios are also very common. Virtually all new cars these days come with a sophisticated radio already fitted. However, these are not the only types of sets to be found in common use. Portable phones are increasing in use, and these contain a receiver. There is also a wide range of commercial uses of radio receivers. They are used in communications systems ranging from short range VHF or UHF walkie talkies to the longer range systems needed for aircraft or maritime communications. Whatever the use of the receiver, the same basic principles apply and the same basic building blocks are used.

A receiver performs two main functions. The first is to remove the modulation from the radio frequency signal. This process, called demodulation or detection, gives signals at audio frequencies which can be amplified by an audio frequency amplifier and passed into headphones or a loudspeaker. The other is to provide selectivity. With the vast number of stations on the radio frequency bands today it is necessary that the receiver is able to tune in the station on the required frequency and reject the others. If this function is not performed well then a number of different stations will be received at one time making it very difficult to copy any of them.

Amplification is another important function which radios provide. The signals picked up by the aerial are very weak, and need to be amplified to a large degree if they are to be heard in a loudspeaker or passed to some other form of unit like a modem where data is extracted.

There are a number of different methods of performing some of these functions. As a result there are a number of different types of format for receivers. Obviously some are much simpler than others. The simplest is a crystal set. This is the most basic type of receiver, and while it is not used for

Figure 5.1 *A typical domestic portable receiver*

any serious listening these days it shows some of the basic principles required for a radio receiver.

Crystal set

The crystal set has been in existence for very many years, often being built by hobbyists these days as a first construction project. The circuit of a simple set is shown in Figure 5.2, and from this it can be seen that very few components are used. Naturally the simplicity reflects the level of performance which can be achieved.

In the circuit $L1$ and VC_1 act together to form a resonant circuit. This performs the function of accepting signals on and around the wanted one and rejecting the others. As the inductor and variable capacitor form a parallel tuned circuit, the impedance reaches a maximum at resonance, as described in Chapter 3. This means that signals on frequencies either side of the resonant frequency pass to earth. At resonance the signals do not pass through the tuned circuit. Instead they pass into the diode. This rectifies the signal so that the amplitude variations from the signal can be obtained. The small smoothing capacitor C_1 acts to remove any remaining radio frequency components and smooths the signal as shown in Figure 5.3.

The rectified signals from the diode can be heard by connecting a pair of headphones to the output of the set. However, it is soon discovered that the

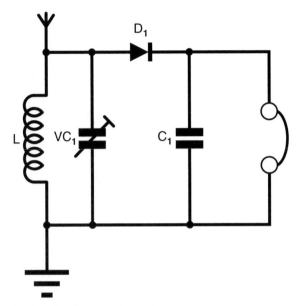

Figure 5.2 *A circuit of a crystal set*

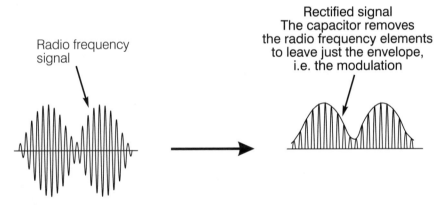

Radio frequency signal

Rectified signal
The capacitor removes
the radio frequency elements
to leave just the envelope,
i.e. the modulation

Figure 5.3 *Demodulating an AM signal*

signal strengths are weak even when a good antenna is used. Also the tuning is very broad and a single strong station may be heard over a large part of the tuning range.

TRF receiver

If a crystal set is built, its limitations very quickly become obvious. The low signal strengths can be overcome by adding amplification. This can be placed after the detector as an audio amplifier, but it can also be added in the radio frequency stages prior to the detector. A receiver of this nature is called a tuned radio frequency (TRF) receiver because the tuning takes place at the radio frequency. This distinguishes it from other types mentioned later.

The lack of selectivity in the crystal set is mainly caused by the fact that there is only one tuning circuit. To overcome this problem and increase the gain still further a circuit known as a regenerative detector can be used. This type of circuit was very popular in the early days of radio when the number of valves had to be kept to a minimum to keep the cost of circuits down to what people could afford.

A regenerative detector works by feeding back some of the signal from the output to the input. A control called a regeneration or reaction control is used to keep the circuit just below the point of oscillation. At this point the maximum usable gain is achieved. The other advantage of this type of circuit is that it also improves the response of the tuned circuit, making it sharper and improving its Q or quality factor.

If the regeneration control is advanced so that the circuit oscillates the receiver is then able to resolve Morse and single sideband signals. The oscillation in the detector beats with a Morse signal to give the characteristic tone. For a single sideband signal it replaces the carrier so that the original audio can be recovered at the detector.

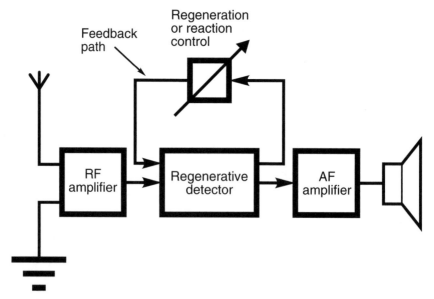

Figure 5.4 *Block diagram of a tuned radio frequency receiver with regeneration*

If a regenerative detector is used then it should be preceded by a radio frequency amplifier. This isolates the oscillation generated in the detector from the antenna. If no RF amplification is present then the oscillation can be radiated or transmitted from the aerial and it can cause interference to others nearby.

Direct conversion receiver

One type of radio which has gained popularity over the years and is now widely used as a simple yet effective solution for applications like amateur radio is the direct conversion set. As its name suggests it uses a process which directly converts the radio frequency signal down to audio frequencies for amplification.

The basic process used in the direct conversion receiver is a process called mixing. This is not like audio mixing where several signals are added together in a linear fashion to give several sounds together. Radio frequency mixing is a non-linear process which involves the level of one signal affecting the level of the other at the output. This process involves the two signal levels multiplying together at any given instant in time.

When two signals are mixed or multiplied in this fashion it is found that other signals are generated. These occur at frequencies equal to the sum and difference of the two original signals as shown in Figure 5.5. In other words if the two input frequencies are f_1 and f_2, then the two resultant signals will appear at $f_1 + f_2$ and $f_1 - f_2$. To give an example if the two original signals are at frequencies of 1 MHz and 0.75 MHz, then the two resultant signals will appear at 1.75 MHz and 0.25 MHz.

If the two frequencies are very near together, for example 1 MHz and 1.001 MHz, then one of the resultant frequencies will appear in the audio part

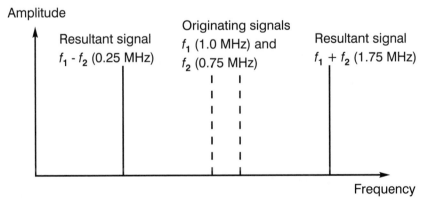

Figure 5.5 *Frequencies produced by mixing two signals together*

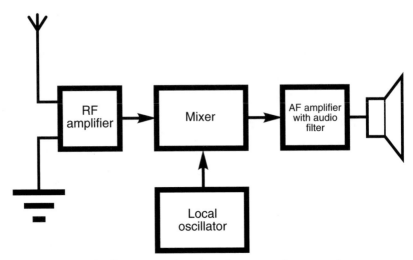

Figure 5.6 *Block diagram of a direct conversion receiver*

of the spectrum and can be amplified by an audio amplifier and passed into a loudspeaker so that it can be heard.

To make up a typical direct conversion receiver a number of circuit blocks are required. These are shown in Figure 5.6. In this the signals first pass into a radio frequency amplifier and tuning stage. This fulfils a number of functions. First of all it amplifies the signals prior to passing them into the mixer. The choice of the level of gain is important, because too much gain can cause the mixer to overload when very strong signals are present. The tuning is also important. If it was not present then signals from a very wide range of frequency would appear at the input to the mixer and this again might cause the mixer to overload. This is particularly important where very strong signals may appear at frequencies far away from the ones being used. Finally the presence of the radio frequency amplifier helps prevent the local oscillator signal from reaching the antenna. The amplifier acts as another stage of isolation reducing it to an acceptable level.

The local oscillator circuit is crucial to the performance of the radio. It must be capable of tuning over the range which the radio is required to cover. In some instances this may only be a small band as in the case of a set designed to cover a single amateur band. At other times the set will need to cover a much wider band of frequencies. In either case the design and construction of the oscillator is crucial. One of the prime requirements is that it should not drift. If it does then the set will need to be re-tuned at intervals if the same station is required.

Once through the mixer the signals enter the audio amplifier. This section may require some filtering. Off-channel signals will produce high frequency interference. This can be reduced by employing an audio filter. For

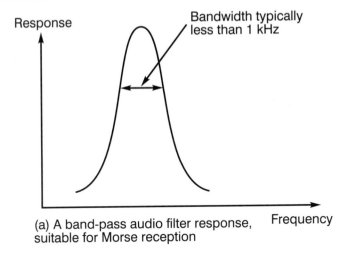

(a) A band-pass audio filter response, suitable for Morse reception

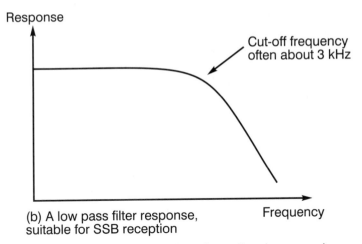

(b) A low pass filter response, suitable for SSB reception

Figure 5.7 *Band-pass, low pass filters in a direct conversion receiver*

transmissions like Morse a narrow band-pass type of filter with a response like that shown in Figure 5.7(a) can be employed. As the name suggests this will just allow through a narrow band of signals and is particularly useful for removing other signals which may appear in the audio band.

For transmissions like SSB a low pass filter with a response like that shown in Figure 5.7(b) is generally used. This allows through the low frequency signals and removes those above a certain cut-off frequency. This will help reduce interference from signals off-channel which produce high frequency audio interference.

Direct conversion receivers can only receive a limited number of types of transmission satisfactorily. Take the example of a Morse transmission. This consists of a carrier wave being turned on and off. When the oscillator in the receiver moves close to the frequency of the Morse signal, an audio note equal in frequency to the difference between the two signals will be generated giving the characteristic on/off tone for Morse. Similarly for single sideband, the oscillator will beat with the sideband to give the required audio. However, for the pitch of the audio to be correct the local oscillator must be on the same frequency that the carrier would have been if it was present.

An amplitude modulated signal is not quite so easy to resolve. If there is a difference in frequency between the carrier of the incoming signal and the local oscillator then an annoying beat note is generated. To be able to resolve the signal satisfactorily the local oscillator must be tuned to exactly the same frequency as the carrier of the required signal. If this is done the audio is recovered and can be heard satisfactorily.

One mode which cannot be resolved is frequency modulation. This has meant that this type of receiver is generally only used on the HF bands where Morse and SSB are widely used.

One of the problems with a direct conversion receiver is called the audio image and it can be demonstrated as the set is tuned through a steady signal. First a high frequency audio heterodyne or oscillation is heard. As the receiver is tuned closer to the signal this note falls until eventually the local oscillator of the receiver and the signal are on the same frequency. At this point the heterodyne frequency reaches zero. This point is known as zero beat. As the receiver continues to be tuned in the same direction the audio note starts to rise in frequency as the local oscillator starts to move away on the other side of the signal. This means that there are two points where a particular audio frequency is obtained. This audio image cannot be removed without the addition of more electronics in the mixer. As the requirements for this circuitry are normally quite exacting most direct conversion sets live with the problem.

The superhet

To overcome the limitations of these types of radio already described, another type of set evolved and gained popularity in the 1920s and 1930s. Called the superhet, which is short for supersonic heterodyne, this type of set is in almost universal use today. It operates by changing the frequency of the incoming signal to a fixed frequency intermediate stage where it can be more easily filtered and amplified before being demodulated to give an audio signal which is amplified in the normal way.

Using the mixing or multiplying process it is possible to change the frequency of a signal up or down as shown in Figure 5.8. To illustrate this, take the example of a signal at 2.0 MHz. This mixes with the local oscillator

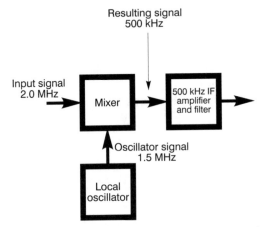

Figure 5.8 *Using a mixer to change the frequency of a signal*

at 1.5 MHz to give a signal at the intermediate frequency of 0.5 MHz or 500 kHz. This signal represents the difference mix product. The sum appears at 3.5 MHz and is easily filtered out as it is easily removed from the frequency of the filters.

Unfortunately there is another mix product which gives a signal that can also pass through the intermediate frequency stages. The first signal represented the difference between the incoming signal minus the local oscillator. However, a signal representing the local oscillator minus the incoming signal can also pass through the filters. Take the same example again where the local oscillator frequency is 1.5 MHz. An incoming signal of 1.0 MHz also gives an output at 500 kHz. It can be seen that signals which can enter the IF stages are at frequencies equal to the local oscillator

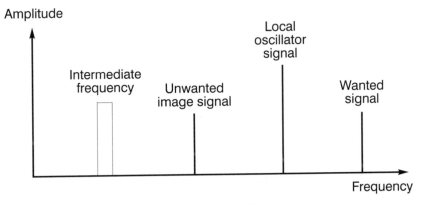

Figure 5.9 *Signals at LO ± IF can enter the IF stages*

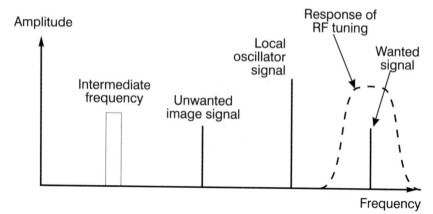

Figure 5.10 *A tuned circuit used to reject the unwanted image signal*

frequency plus the intermediate frequency and the local oscillator minus the intermediate frequency as shown in Figure 5.9.

It is obviously undesirable to have two signals on totally different input frequencies which are able to pass through the intermediate frequency filters and amplifiers. Fortunately it is relatively easy to remove the unwanted or image signal to leave only the required one. This is achieved by placing a tuned circuit in the radio frequency stages to remove the signal as shown in Figure 5.10.

From the diagram it can be seen that this tuned circuit does not need to reject signals on adjacent channels like the main intermediate frequency filters. It is only necessary to reject the image signal. This happens to fall at a frequency equal to twice the intermediate frequency away from the wanted signal. In the case of the example this is 1 MHz away from the wanted signal. This is compared to stations on adjacent channels which are rejected by the main intermediate frequency filters which may be a few kilohertz away, or in some cases less.

Tuning the receiver is primarily accomplished by changing the frequency of the local oscillator. If it is moved up in frequency by 100 kHz then the frequency of the signals which are received will be 100 kHz higher. In the example used earlier, if the local oscillator is 100 kHz higher at 1.6 MHz then signals will be received at 2.1 MHz, assuming the higher frequency signals are wanted and those at 1.1 MHz are rejected as the image.

If the local oscillator and hence the receiver frequency is varied, it is necessary to ensure that the RF tuning also moves at the same rate. In this way the RF tuning will be set to the correct frequency and this will ensure the wanted signal passes through without being attenuated. It also ensures that the image signal is fully rejected.

Figure 5.11 *A ganged tuning capacitor*

To ensure that this occurs the tuning for the local oscillator and RF circuits must be linked so that they track together. In many older sets ganged tuning capacitors were used like that shown in Figure 5.11. Here the two sections of the capacitor are mechanically linked. In this way the two tuned circuits can be varied by the same degree at the same time. In more modern receivers electronic methods of tuning are used, and these are designed to track at the same rate to maintain the correct circuit conditions.

Basic superhet receiver

Having looked at the superhet principle the various blocks can be put together to form a complete radio, as shown in Figure 5.12. The signals enter the radio frequency circuits where the required band of frequencies is selected and those which might give rise to image signals are rejected. In the same stages the signals are amplified to the required level and passed into the mixer or multiplier circuit. Here they are mixed with the output from the local oscillator to convert them to the intermediate frequency. As the mixer performs the action of changing the frequency of the signals it is sometimes called a frequency changer.

In many portable sets the RF tuning, oscillator and mixer functions are performed by a single transistor. These circuits are very cost effective to produce, requiring far fewer components than if the circuit was made up by

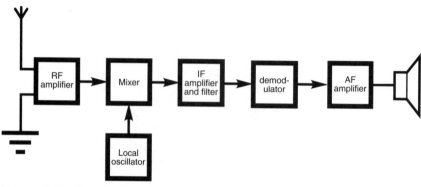

Figure 5.12 *Block diagram of a basic superhet*

Figure 5.13 *A hi-fidelity tuner which uses the superhet principle (Courtesy of Sony)*

separate circuit blocks. Naturally the performance of the circuit cannot be optimized to give the best performance, but it is normally quite adequate for many of the lower cost sets available on the market.

For higher performance sets high grade mixers are used. There are a number of parameters which are of importance when considering mixer performance. The first is the conversion gain or loss. This is the ratio of the signal level at the RF signal input to the output at the intermediate frequency port. The noise performance is also important as will be described later. As mixers are often located near the front end stages of the set they have a large bearing on the sensitivity and noise performance of the whole set. The isolation between ports is also specified for many ready-made mixers. This figure represents the amount of signal from one port which finds its way onto another.

A number of circuits can be used as mixers. Diode ring mixers are popular in many high performance applications. Although they give a conversion loss, typically of around 7 dB, other parameters in their performance are normally very high.

Active mixers have a conversion gain. Of these, the circuits based around FETs generally give the best performance because of the characteristics of the devices. They give lower levels of unwanted or spurious signals. Often

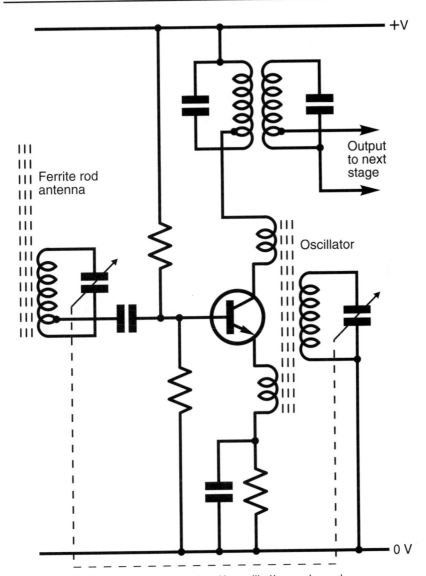

Figure 5.14 *Circuit of a typical self-oscillating mixer stage*

the circuits used what are called double balanced circuits, but those using a single dual gate MOSFET, like that shown in Figure 5.16, perform well and are used in many hi-fi tuners and other radio receivers. In these circuits the signal is applied to one of the gates and the local oscillator or switching signal is applied to the other. The output is taken from the drain or source in the usual way.

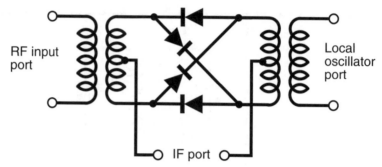

Figure 5.15 *A high performance diode mixer*

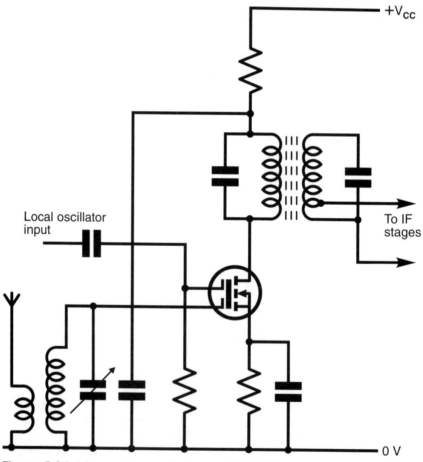

Figure 5.16 *Mixer using a dual gate MOSFET*

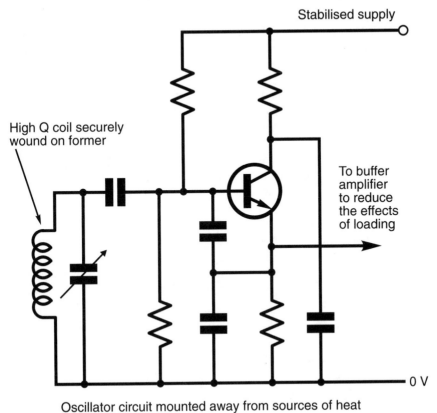

Oscillator circuit mounted away from sources of heat

Figure 5.17 *An LC tuned variable frequency oscillator*

The performance of the local oscillator is very important, particularly where frequency drift is concerned. Today frequency synthesizers are often used. As they normally have a crystal oscillator or crystal oven as their reference, their stability is usually very good. Many older receivers have LC tuned variable frequency oscillators like that shown in Figure 5.17. These oscillators are prone to drift in frequency. This is generally caused by changes in temperature causing the values of some of the components to alter slightly. Even very small changes will give noticeable amounts of drift. If the oscillator drifts then the set will need to be re-tuned. This is not usually a problem from long and medium wave sets, but where frequencies are higher and modes of transmission like SSB are used the stability of the oscillator is of great importance.

To overcome the effects of drift a number of precautions must be employed. The circuit should be run from a stabilized power supply. Changes in supply voltage cause the circuit parameters of some components to change. This is particularly true of some semiconductor devices like bipolar transistors. Coils should be wound as securely as possible and the tuned circuit should not be switched if at all possible. In addition to this the tuned circuit should be designed to have a high Q. Finally the whole oscillator circuit must be kept away from any sources of heat.

After the mixer the signals reach the IF stages. It is here that the majority of the gain and selectivity are provided. Most of today's sets utilize integrated circuits as the basic building blocks, and the actual circuits depend upon the particular IC being used. Some radios still use discrete components and a typical circuit is shown in Figure 5.18. Often two stages are used in series to give the required gain and selectivity. In this circuit the selectivity is provided by the coupling transformer. This is tuned using a ferrite adjuster and each one requires adjustment to the correct frequency.

A number of standard intermediate frequencies are used. Frequencies of 455 or 465 kHz are widely used for medium and long wave broadcast radios.

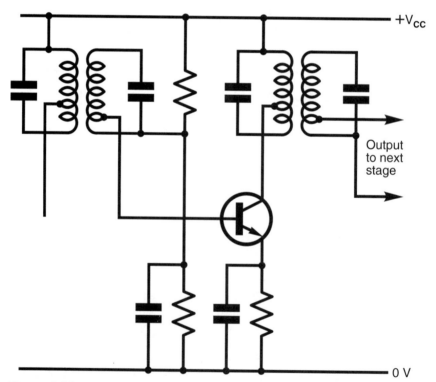

Figure 5.18 *A typical IF amplifier stage using discrete components*

The low frequency means that it is possible to obtain sufficient selectivity with a minimum number of stages.

Other frequencies are used for other purposes. For VHF FM receivers 10.7 MHz is a standard, and this frequency is also used in many hand-held transceivers for mobile or on-site communications. Frequencies of 1.6 and 9 MHz are also used in some high performance short wave communications receivers.

Signals leaving the IF stages are still in the form of a radio frequency signal where modulation is superimposed on a carrier. This modulation needs to be removed by a circuit called a demodulator so that the audio signal can be amplified by an audio amplifier before being converted into sound by a loudspeaker or headphones.

Overcoming image response problems

The problem of receiving an image signal is one of the major drawbacks of the superhet receiver. Sufficient selectivity needs to be placed into the front end circuits to be able to reduce any unwanted image signals to a sufficiently low level. Unfortunately a receiver covering a wide range of frequencies will have a much greater problem at the top end of the frequency range than at the bottom.

The reason can be deduced from looking at the Q of the RF circuits. Tuning is accomplished by changing the capacitance in the circuit and this means that the Q remains constant. This results in the bandwidth increasing in line with the frequency. As the frequency doubles so does the bandwidth. As the difference between the wanted signal and the image remains constant at twice the intermediate frequency it means that the RF tuning will reject the image signal much less at higher frequencies, as shown in Figure 5.19. At higher frequencies the frequency difference becomes a much smaller proportion of the frequency of operation.

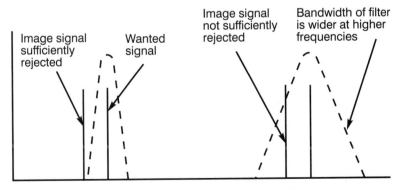

Figure 5.19 *Image response is degraded at higher frequencies*

There are two main ways of improving the image response. The first is to make the RF tuning much sharper. If this is done then it is necessary to ensure that the RF tuning tracks the local oscillator very accurately. If the improved selectivity of the front end stages is achieved by using more than one tuned circuit, then these have to track one another as well as the local oscillator. Any misalignment of the circuits will cause the wanted signal to be reduced in strength and the image to increase. This can be difficult and for this reason multiple tracked tuned circuits are rarely used.

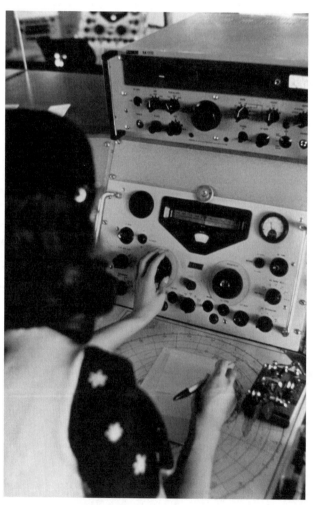

Figure 5.20 *A high performance valve communications receiver (Courtesy of Racal Radio)*

A more satisfactory option is to increase the intermediate frequency. By doing this the frequency difference between the wanted signal and image is increased. This is the solution that is normally adopted these days because other various techniques are available which allow higher intermediate frequencies to be used.

As image response is an important parameter in performance radio receivers it is a parameter which is often specified, particularly for high performance communications receivers. It is defined as a given number of decibels at a particular frequency. For example, it may be 50 dB at 30 MHz. This indicates that if signals of the same strength on the image and wanted frequencies were present at the input, then the image one would be 50 dB lower than the wanted one at the output of the receiver. The frequency at which the measurement is made has to be included because the figure varies with the frequency of operation. A typical figure for a modern short wave communications receiver would be 80 dB at most frequencies in the short wave spectrum falling to about 60 dB at the worst frequencies.

IF breakthrough

Another problem which can occur happens when signals from the aerial pass across the RF sections and enter the IF directly. Normally intermediate frequencies are chosen to fall at frequencies which are not used by high powered stations. Although this alleviates the problem to a large extent, there are sometimes situations where this is not feasible.

To ensure that breakthrough is not a problem care must be taken during the design of the receiver to ensure that there is sufficient isolation between the signal input and the IF stages. In some sets special filters are included at the aerial to ensure that signals from the aerial at the intermediate frequency are removed as they enter the set.

Figures for IF breakthrough are quoted in the same way as they are for image rejection. In the case of most high performance receivers it is possible to achieve figures of more than 60 dB rejection. In some cases figures of 100 dB have been quoted.

Spurious signals

Virtually all receivers generate signals themselves, many of which can be picked up by the set and received as if they had entered the set through the aerial socket. In many cases these signals go unnoticed because they are masked out by real signals, but in some instances they act as unwanted interference degrading the performance of the set.

Spurious signals can be generated in a number of ways. Most of today's receivers use a variety of oscillators in their circuitry. It is very easy for these signals, or their harmonics, to enter the signal path and appear as signals.

On wide-band receivers it is often impossible to eliminate all the spurious signals, although most of them can be reduced to levels where they do not cause a major problem. Some receivers will quote known spurious signals.

For narrow-band receivers like those in mobile phones it is possible to design the set so that no significant spurious signals are generated within the frequency range of the receiver.

Automatic gain control

A circuit which is standard on all superhet radios these days is called an automatic gain control (AGC). Occasionally this circuit is called an automatic volume control (AVC) although this term is not particularly common these days and the more correct description of AGC is used.

This circuit is used to help compensate for the enormous variation in signal levels which are encountered. From one station to the next there could be a difference of 90 dB or more. Even when the receiver is tuned to one station the signal may still vary. One example of this occurs with car radios when the vehicle is travelling, and hills and other obstructions mean that the signal strength is always varying. Signal strengths on the short wave bands also vary considerably and the AGC can be used to compensate for this.

The AGC operates by sensing the level of the signal when it is demodulated. The voltage generated is used to adjust the gain of some of the earlier stages. The greater the level of the signal being received, the greater the voltage produced. This enables the gain of the earlier stages to be reduced to a greater extent, thereby keeping the demodulated signal almost constant.

All portable broadcast receivers have AGC circuits which utilize a minimum number of components. However, for more sophisticated receivers like communications sets the design of the AGC system can become considerably more complicated.

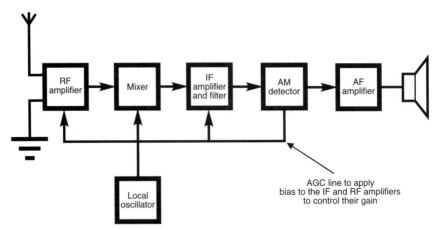

Figure 5.21 *An AGC system in a superhet radio*

The first requirement of the system is that it should not reduce the gain of the set until the signal has achieved a sufficiently high signal to noise ratio. Time constants also have to be taken into consideration. For an amplitude modulated signal it is essential that the AGC system does not act so quickly that the modulation is removed from the carrier. To ensure that this does not happen a simple filter is normally used. This may be a simple capacitor resistor network, giving a time constant of about a quarter of a second.

When transmissions such as Morse and single sideband are used different time constants are required. When using AM the carrier is present all the time, but this is not so with either Morse or single sideband. For single sideband the levels of the signal varies according to level of audio present at the transmitter, and it falls to zero when no sound is present. For an AGC system to be able to cope adequately with this type of transmission a dual time constant is needed. It should react very quickly when a new signal appears. In other words it must have a fast attack time. Typically figures of less than 0.1 seconds are used. The decay time must then take account of gaps in the speech or differing levels between syllables in words. Often decay times of a second or more are employed to prevent the AGC level varying unduly and increasing operator fatigue.

For Morse transmissions a fast attack time is again required, but a faster decay time is generally employed as the pauses in transmission are not usually so long and the power level when the key is down is constant. The AGC only has to take account of the breaks between the dots and dashes which are generally much less than half a second.

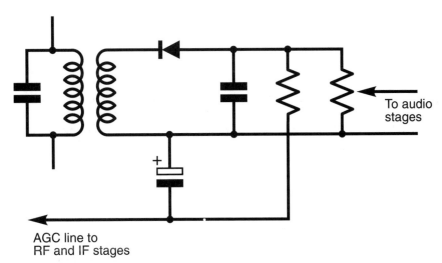

AGC line to
RF and IF stages

Figure 5.22 *A simple detector giving a voltage output for automatic gain control*

To generate the voltage for the AGC a simple diode detector may be used. One like that shown in Figure 5.22 is often used, especially in medium wave broadcast sets. This performs the normal detection or demodulation function for the audio, but also generates a voltage which is filtered before being applied to the earlier stages of the set to control the gain. In more sophisticated sets a separate detector may be employed. This is likely to have different time constants for each mode being used. In many cases this may be linked to the mode switch on the set, or there may be a separate control on the front panel.

Multiple conversion sets

The idea of the superhet radio has many advantages over the other types of set which were previously used. It gives better selectivity, it is possible to easily include several types of demodulator which can be switched in as required, it is capable of being made very sensitive, and it can have a whole host of facilities included into its design. However, in its basic form it does have a number of disadvantages. In this first instance its image rejection may be poor, especially at high frequencies as previously described. If the intermediate frequency is increased to improve this then the selectivity may suffer. A further problem is the stability of the set. To tune the receiver the local oscillator has to be made variable. In turn this means that the receiver is prone to drift once it has been set to a particular frequency. This can be particularly annoying when the set has to be regularly adjusted to keep it tuned to a station. This was a common problem with the older valve receivers. One of the major causes of drift arises from changes in temperature. As valves themselves generated large amounts of heat, these sets often drifted badly. Only after they had been turned on for a few hours would their temperature stabilize and the level of drift reduce.

To overcome some of these problems multiple conversion receivers may be used. By using more than one conversion the signal may be stepped down to the final IF in two stages. The most obvious method of achieving this is to adopt a system as shown in Figure 5.23. Here a variable frequency oscillator is used for the first conversion and a crystal oscillator for the second conversion. In this scheme the first conversion takes the signal down to a higher intermediate frequency to increase the difference between the wanted signal and the image. A second conversion using a fixed frequency oscillator is used to convert the signal down to the second intermediate frequency stage. This oscillator is on a fixed frequency and is usually based around a quartz crystal for convenience and stability.

As the first intermediate frequency is at a higher frequency than before, a major improvement in image response is seen. Unfortunately the problem of drift still remains as the variable frequency oscillator still operates at a high frequency.

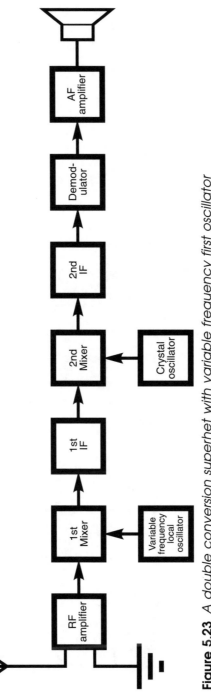

Figure 5.23 *A double conversion superhet with variable frequency first oscillator*

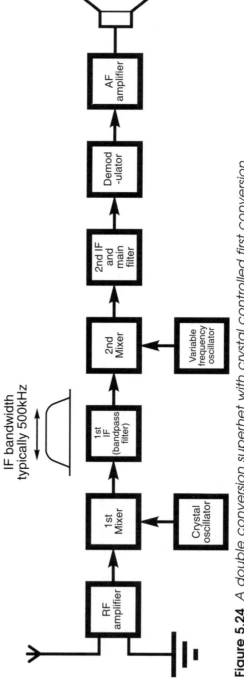

Figure 5.24 *A double conversion superhet with crystal controlled first conversion*

One method of overcoming the drift problem is to use a crystal controlled first conversion as shown in Figure 5.24. As the first oscillator is crystal controlled it is very much more stable. The first conversion is used to convert a band of frequencies down to the first IF which has a relatively wide bandwidth, possibly as much as 500 kHz. A variable frequency oscillator running at a much lower frequency is used for the second conversion. This type of set also has the advantage that the variable frequency oscillator is not switched to cover the different bands as it was before. This also improves its stability.

The major drawback of this type of design is that it needs a large number of crystals in the first conversion if it is to cover a wide spread of frequencies. However, this solution was very successfully used for some communications receivers. Another solution to the problem of drift is to use a much more stable form of local oscillator, and this is the approach which is adopted in most sets today.

Synthesizers

Many receivers using frequency synthesizers boast such terms as 'PLL', 'Quartz', and 'Synthesized' in their specifications or advertising literature. Frequency synthesizers are found in most of today's receivers. They are able to offer very high degrees of stability, they can operate over a wide range of frequencies, and they easily interface to microprocessor circuitry, enabling them to be controlled by these circuits so that the set can provide new degrees of flexibility and facilities. Push-button tuning, frequency memories, scanning and a whole host of other facilities are now available in many sets as a result of the combination of microprocessors and frequency synthesizers.

A frequency synthesizer is based around a phase locked loop which uses the idea of phase comparison for its operation. The phase of a signal is the position within a cycle. Often this is likened to the point on the signal travelling around a circle with a complete cycle being equal to 360°. Each point on the sine wave has an equivalent point on the circle as shown in Figure 5.25. If there are two signals they may not be at the same point in the cycle and there is a phase difference between them. The phase angle between them is equal to the angular difference on the circle as shown.

From the block diagram of a basic loop shown in Figure 5.26 it can be seen that there are three basic circuit blocks, a phase comparator, voltage controlled oscillator, and loop filter. A reference oscillator is sometimes included in the block diagram, although this is not strictly part of the basic loop even though a reference signal is required for its operation.

The loop operates by comparing the phase of two signals. The signals from the voltage controlled oscillator and reference enter the phase comparator and in this circuit a third signal equal to the phase difference between the two is

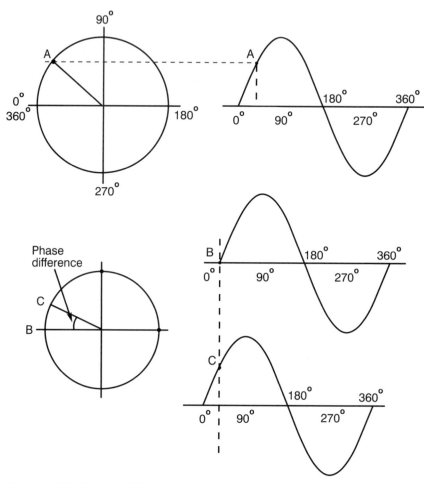

Figure 5.25 *Phase difference between two signals*

produced and is passed through the loop filter. This performs a number of functions within the loop, but in essence it removes any unwanted products from the error voltage before the tune voltage is applied to the control terminal of the voltage controlled oscillator. The error voltage is such that it tries to reduce the error between the two signals entering the phase comparator. This means that the voltage controlled oscillator will be pulled towards the frequency of the reference, and when in lock there is a steady state error voltage. This is proportional to the phase error between the two signals, and is constant. Only when the phase between two signals is changing is there a frequency difference. As there is no phase change when

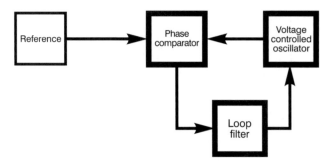

Figure 5.26 *Block diagram of a basic phase locked loop*

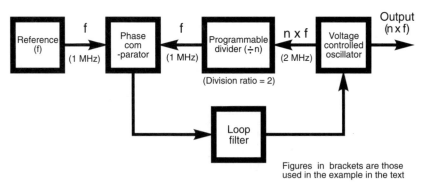

Figures in brackets are those
used in the example in the text

Figure 5.27 *A programmable divider added into a phase locked loop enables the frequency to be changed*

the loop is in lock this means that the frequency of the voltage controlled oscillator is exactly the same as the reference.

A phase locked loop needs some additional circuitry if it is to be converted into a frequency synthesizer. A frequency divider is added into the loop between the voltage controlled oscillator and the phase comparator as shown in Figure 5.27. The divider divides the frequency of the incoming signal by a certain ratio. If the divide ratio is set to three, then the output frequency is a third of the input, and so on.

Programmable dividers are widely used in a variety of applications, including many radio frequency uses. Essentially they take in a pulse train like that shown in Figure 5.28, and give out a slower train. In a divide-by-two circuit only one pulse is given out for every two that are fed in and so on. Some are fixed, having only one division ratio. Others are programmable and digital or logic information can be fed into them to set the division ratio.

When the divider is added into the circuit the loop still tries to reduce the phase difference between the two signals entering the phase comparator.

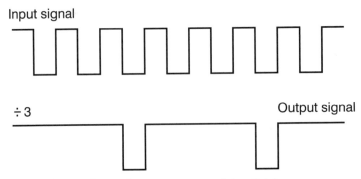

Figure 5.28 *Operation of a programmable divider*

Again when the circuit is in lock both signals entering the phase comparator are exactly the same in frequency. For this to be true the voltage controlled oscillator must be running at a frequency equal to the phase comparison frequency times the division ratio as shown in the diagram.

Take the example when the divider is set to two, and a reference frequency of 1 MHz. When the loop is in lock the frequency entering both ports of the phase comparator will be 1 MHz. If the frequency divider is set to two this means that the frequency entering the divider must be 2 MHz, and the voltage controlled oscillator must be operating at this frequency. If the divider is changed to divide by three, then this means that the voltage controlled oscillator will need to run at 3 MHz and so on. From this it can be seen that the loop will increment by a frequency step equal to the comparison frequency when the division ratio is increased by one. In other words the step frequency is equal to the comparison frequency.

As the output frequency from the loop is locked to the reference signal, its stability is also governed by this. Accordingly most reference oscillators use quartz crystals to determine their frequency. In this way any drift is minimized, and the optimum accuracy is obtained.

Most synthesizers need to be able to step in much smaller increments if they are to be of any use. This means that the comparison frequency must be reduced. This is usually accomplished by running the reference oscillator at a frequency of a megahertz or so, and then dividing this signal down to the required frequency using a fixed divider. In this way a low comparison frequency can be achieved. This gives the block diagram of a basic synthesizer shown in Figure 5.29.

Placing a digital divider is not the only method of making a synthesizer using a phase locked loop. It is also possible to use a mixer in the loop as shown in Figure 5.30. By using a mixer in the loop, an offset to the reference frequency is generated.

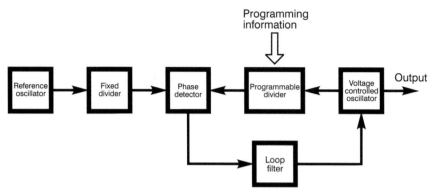

Figure 5.29 *Comparison frequency reduced by adding a fixed divider after the reference oscillator*

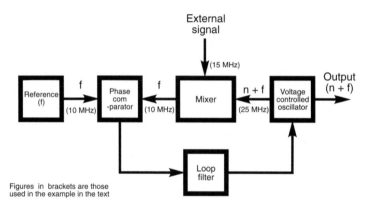

Figures in brackets are those
used in the example in the text

Figure 5.30 *A phase locked loop with mixer*

The way in which the loop operates with the mixer incorporated can be analysed in the same manner that was used for the loop with a divider. When the loop is in lock the signals entering the phase detector are at exactly the same frequencies. The mixer adds an offset equal to the frequency of the signal entering the other port of the mixer. To illustrate the way this operates figures have been included. If the reference oscillator is operating at a frequency of 10 MHz and the external signal is at 15 MHz then the VCO must operate at either 5 MHz or 25 MHz. Normally the loop is set up so that the mixer changes the frequency down and if this is the case then the oscillator will be operating at 25 MHz.

It can be seen that there may be problems with the possibility of two mix products being able to give the correct phase comparison frequency. It happens that as a result of the phasing in the loop, only one will enable it to lock. However, to prevent the loop getting into an unwanted state the range of the VCO is limited. For loops which need to operate over a wide range a steering voltage is added to the main tune voltage so that the frequency of the loop is steered into the correct region for required conditions. It is relatively easy to generate a steering voltage by using digital information from a microprocessor and converting this into an analogue voltage using a digital to analogue converter (DAC). The fine tune voltage required to pull the loop into lock is provided by the loop in the normal way.

Multi-loop synthesizers

Many high performance synthesizers use a number of loops, and use both mixers and digital dividers to obtain the required output. By combining a number of loops it is possible to produce high class wide range oscillators with low levels of phase noise and very small step sizes. If a single loop is used then there are many shortfalls in the performance.

There is a large variety of ways in which multi-loop synthesizers can be made, dependent upon the requirements of the individual system. However, as an illustration a two loop system is shown in Figure 5.31. This uses one loop to give the smaller steps and the second provides larger steps. This principle can be expanded to give wider ranges and smaller steps.

The operation of the synthesizer is not complicated. The first loop has a digital divider. This operates over the range 19 to 28 MHz and has a reference frequency of 1 MHz to provide steps of 1 MHz. The signal from this loop is fed into the mixer of the second one. The second loop has division ratios of 10 to 19, but as the reference frequency has been divided by 10 to 100 kHz this gives smaller steps.

The operation of the whole loop can be examined by looking at extremes of the frequency range. With the first loop set to its lowest value the divider is set to 19 and the output from the loop is at 19 MHz. This feeds into the second loop. Again this is set to the minimum value and the frequency after the mixer must be at 1.0 MHz. With the input from the first loop at 19 MHz this means that the VCO must operate at 20 MHz if the loop is to remain in lock.

At the other end of the range the divider of the first loop is set to 28, giving a frequency of 28 MHz. The second loop has the divider set to 19, giving a frequency of 1.9 MHz between the mixer and divider. In turn this means that the frequency of the VCO must operate at 29.9 MHz. As the loops can be stepped independently it means that the whole synthesizer can move in steps of 100 kHz between the two extremes of frequency. As mentioned before this principle can be extended to give greater ranges and smaller steps, providing for the needs of modern receivers.

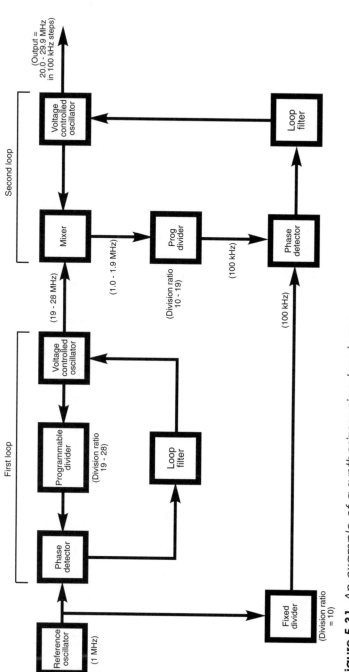

Figure 5.31 *An example of a synthesizer using two loops*

Direct digital synthesizers

Although phase locked loops form the basis of most synthesizers, other techniques can be used. One which is gaining popularity is called the direct digital synthesizer (DDS). Although the techniques behind this method of generating signals have been known for many years, it was necessary for semiconductor technology to develop to a sufficient level before they became a viable option. In fact many synthesizers combine both PLL and DDS techniques to gain the best of both systems.

As the name suggests this method generates the waveform directly from digital information. This is in contrast to a phase locked loop system which uses a signal on one frequency, i.e. the reference, to generate a signal on the required frequency by the action of the loop.

The synthesizer operates by storing various points in the waveform in digital form and then recalling them to generate the waveform. Its operation can be explained in more detail by considering the phase advances around a circle as shown in Figure 5.32. As the phase advances around the circle this corresponds to advances in the waveform as shown. The idea of advancing phase is crucial to the operation of the synthesizer as one of the circuit blocks is called a phase accumulator. This is basically a form of counter. When it is clocked it adds a preset number to the one already held. When it fills up, it resets and starts counting from zero again. In other words this corresponds to reaching one complete circle on the phase diagram and restarting again.

Once the phase has been determined it is necessary to convert this into a digital representation of the waveform using a waveform map. This is a memory which stores a number corresponding to the voltage required for each value of phase on the waveform. In the case of a synthesizer of this nature it is a sine lookup table, as a sine wave is required. In most cases the

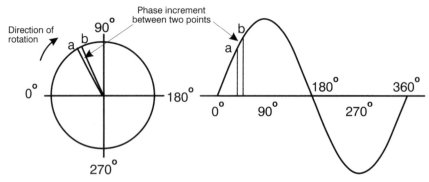

Figure 5.32 *Operation of the phase accumulator in a direct digital synthesizer*

Frequency information
(phase increment)

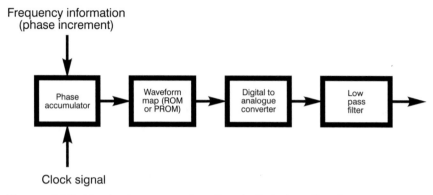

Clock signal

Figure 5.33 *Block diagram of a direct digital synthesizer*

memory is either a read only memory (ROM) or programmable read only memory (PROM).

The next stage in the process is to convert the digital numbers into an analogue voltage using a digital to analogue converter (DAC). This signal is filtered to remove any unwanted signals and amplified to give the required level as necessary.

There are a number of spurious signals which are generated by a direct digital synthesizer. The most important of these is one called an alias signal. Here images of the signal are generated on either side of the clock frequency and its multiples. For example, if the required signal had a frequency of 3 MHz and the clock was at 10 MHz then alias signals would appear at 7 MHz and 13 MHz as well as 17 MHz and 23 MHz, etc. These can be removed by the use of a low pass filter.

Tuning is accomplished by varying the size of the step or phase increment between different sample points. By doing this the speed at which the system advances around the cycle is increased or decreased. This can be done because there will be many more points stored in the waveform map than it is necessary to regenerate the signal. For example, one frequency may be generated by adding 5067 to the value in the phase accumulator each time, whereas the next frequency will be obtained by adding 5068. All the time the clock remains the same, and it is normally a crystal oscillator which may be oven controlled to ensure the frequency stability of the whole system.

From this it can be seen that there is a finite difference between one frequency and the next, and that the minimum frequency difference or frequency resolution is determined by the total number of points available in the phase accumulator. A 24 bit phase accumulator provides just over 16 million points and gives a frequency resolution of about 0.25 Hz when used with a 5 MHz clock. This is more than adequate for most purposes.

It is worth noting that to change frequency the synthesizer simply starts adding the new number onto the total held in the phase accumulator. There is no need for a reset, which means that the synthesizer can change frequency virtually instantaneously. This can be a major advantage of this type of synthesizer in some applications. Phase locked loop systems have a distinct settling time which slows down the rate at which they can change. This may be a critical parameter when receivers are required to scan.

The other advantage of direct digital synthesizers is that they can operate over a very wide range. The major limitation is the top frequency of operation which is governed by the integrated circuits.

These synthesizers do have some disadvantages. They are currently more expensive than their phase locked loop counterparts, but this may change with further developments in semiconductor technology. The other disadvantage is that they produce small spurious signals close to the wanted signal. Normally these can be reduced to acceptable levels, but they may pose a problem in some applications.

Phase noise and reciprocal mixing

One of the main problems with frequency synthesizers is the fact that they can generate high levels of phase noise if care is not taken in the design. This noise is caused by small amounts of phase jitter on the signal, and it manifests itself as noise spreading out either side of the signal as shown in Figure 5.34.

Any signal source will have some phase noise. Crystal oscillators are very good, and free running variable frequency oscillators normally perform well. Unfortunately synthesizers, especially those based around phase locked loops, do not always fare so well and this can adversely affect the performance of the radio in terms of reciprocal mixing.

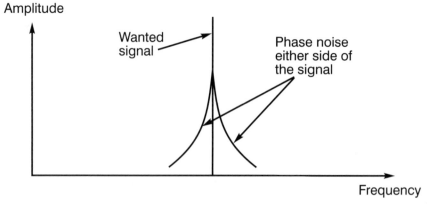

Amplitude

Wanted signal

Phase noise either side of the signal

Frequency

Figure 5.34 *Phase noise on a signal*

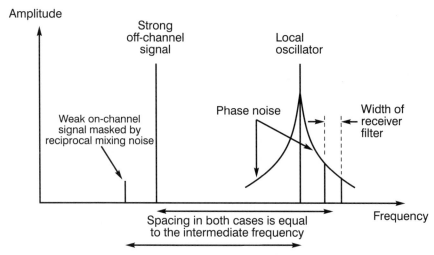

Amplitude

Figure 5.35 *Reciprocal mixing*

This can be explained by considering the receiver when it is tuned to a strong signal. The local oscillator mixes with the wanted station to produce which falls within the pass-band of the receiver filters. If the receiver is tuned off-channel by a given amount, for example 10 kilohertz, then the station can mix with the local oscillator phase noise 10 kilohertz away from the main oscillator signal as shown in Figure 5.35. If the incoming station signal is very strong the reciprocal mixing effect can mask out weaker stations, thereby reducing the effective sensitivity of the set in the presence of strong off-channel signals.

Filters

There is a wide variety of different types of filter found in today's receivers. LC, crystal, mechanical, monolithic, ceramic and audio filters can all be found. Each one has its own advantages and disadvantages, and by careful use of the correct type in particular positions in the set, excellent selectivity can be obtained. With the crowded band conditions experienced today, this is extremely important, and the selectivity of a set will be one of the main features described in its specification.

The basic selectivity of a set is provided in the IF stages. Here the stations on adjacent channels are rejected, and the performance of the filters here will determine the performance of the whole set. Often several filters placed at different points in the set provide the selectivity. However, in most sets one high quality unit is used to provide the majority of the selectivity. In multi-conversion sets this filter is placed at the last intermediate frequency where the frequency is generally the lowest. In this way it is possible to achieve the most effective filtering.

Selectivity and filters

The selectivity of a receiver is of prime importance. Each filter has a finite bandwidth, and this must be taken into consideration when selecting a filter for a receiver. With the number of stations using the radio spectrum today it is necessary to ensure that only signals on the selected frequency or channel are received. As transmissions have a finite bandwidth which depends upon the type of transmission in use it is necessary to match the filter bandwidth to the transmission. If the filter is too wide, as shown in Figure 5.36, then there is the possibility of picking up unwanted signals. On the other hand, if it is too narrow then part of the wanted signal will be rejected and this will result in distortion. It is therefore necessary to use filters of the required bandwidth. Typically medium wave AM transmissions need about 9 kHz, and the standard for short wave AM transmissions is about 6 kHz. For SSB, 2.7 kHz or thereabouts is usually used, and for Morse, bandwidths of 500 or 250 Hz are often used. The VHF FM broadcasts normally require a bandwidth of about 200 kHz to ensure the whole of the signal is received, and other types of transmission all have their own requirements.

Demodulator

The signals which appear at the output of the intermediate frequency amplifier and filter section of a receiver consist of a radio frequency type of signal. In other words they consist of a carrier with the required modulation superimposed. To remove the modulation, a circuit called a demodulator is required. While these circuits are often termed demodulators, the name detector is also used almost interchangeably as well.

The type of demodulator and hence the circuitry of the demodulator will depend upon the type of modulation which is being used. Often different demodulators are required for different modes of transmission. When a receiver needs to be able to demodulate a variety of types of signal, different detectors are switched in.

AM demodulation

The simplest type of detector is a simple diode detector like that used in the crystal set and shown in Figure 5.2. This type of detector is widely used in many receivers, although it can give relatively high levels of distortion for

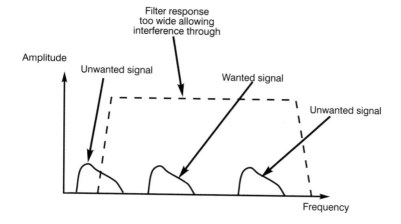

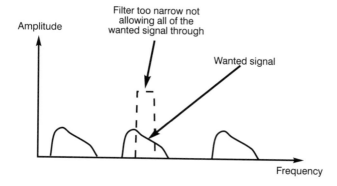

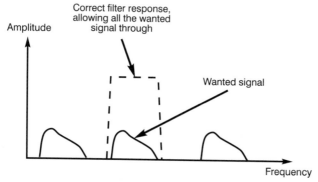

Figure 5.36 *Filter bandwidths must match the transmission bandwidth*

some applications. It is also not particularly good when selective fading occurs on short wave signals. Under these conditions certain of the frequencies of a signal are not received for short periods of time, and this can lead to very high levels of distortion.

SSB and Morse demodulation

To receive Morse and single sideband it is necessary to have additional circuitry. In order to hear the Morse signal it is necessary to beat the incoming signal with an internally generated one so that the typical Morse signal of a tone being turned on and off can be heard. If a normal AM detector was used, then a series of clicks and thumps would be heard as the carrier was turned on and off.

The internal oscillator is generally called a beat frequency oscillator or BFO. The incoming signal and the beat signal are passed into a mixer where they are mixed together. The resulting audio tone is equal in frequency to the difference between the signal and the BFO. As the mixing process is a multiplication process, and the output is the product of the two input signals, this type of detector is often called a product detector.

The same type of detector is also used for receiving single sideband. Here the beat frequency oscillator is used to reintroduce the carrier as shown in Figure 5.37. As a result of this application the oscillator may also be known as the carrier insertion oscillator or CIO.

The frequency of the BFO needs to be set to give the optimum performance. The signal being received should be tuned for the best

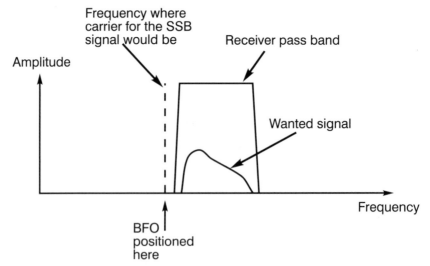

Figure 5.37 *Action of a beat frequency oscillator*

reception, by placing it in the middle of the receiver pass-band. The setting of the oscillator is then adjusted to suit. This may often mean that it is just outside the pass-band of the set. However, this does not matter as it comes after the selectivity.

Synchronous AM demodulation

Improvements in amplitude modulation reception can be achieved by using a system known as synchronous detection. Essentially it uses a beat frequency oscillator and product detector or mixer for AM. For this to work the beat frequency oscillator must be kept on the same frequency or zero beat with the carrier. There are a number of methods of doing this, but the one which is most commonly used employs a high gain limiting amplifier. This takes the AM signal and removes the modulation to leave only the carrier, as shown in Figure 5.38. This is used as the beat frequency oscillation to demodulate the signal.

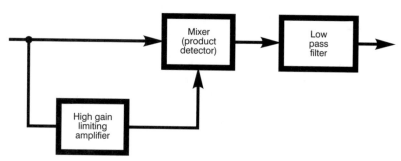

Figure 5.38 *Synchronous detector*

Synchronous detection of AM gives a far more linear method of demodulation. It also gives more immunity against the effects of selective fading on the short wave bands.

FM demodulation

In order to be able to demodulate an FM signal it is necessary to convert the frequency variations into voltage variations. In order to be able to achieve this the circuit of the demodulator needs to be frequency dependent. In most cases the demodulator has a response like that shown in Figure 5.39. For obvious reasons the response curve is known as an 'S' curve. From this it can be seen that as the signal moves up in frequency a higher voltage is produced and vice versa. In this way the frequency variations of the signal are

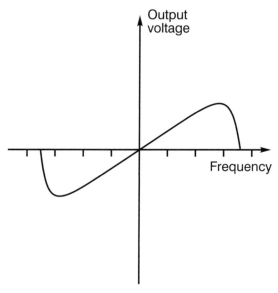

Figure 5.39 *The characteristic curve of an FM demodulator*

converted into voltage variations which can be amplified by an audio amplifier before being passed into headphones or a loudspeaker.

To enable the best detection to take place the signal should be centred about the middle of the curve. If it moves off too far then the characteristic becomes less linear and higher levels of distortion result. Often the linear region is designed to extend well beyond the bandwidth of a signal so that this does not occur. In this way the optimum linearity is achieved. Typically the bandwidth of a circuit for receiving VHF FM broadcasts may be about 1 MHz whereas the signal is only 200 kHz wide.

It is also found that when no signal is present high levels of noise are received. To overcome this most receivers have what is called a squelch circuit. This cuts off the audio when no signal is present. Many scanners and hand-held VHF or UHF transceivers have a control to adjust the signal level below which the audio cuts off. By using this control the set can be adjusted to detect very low level signals if necessary. Hi-fi tuners also incorporate a squelch circuit in most instances. However, they seldom have a control to adjust as it is usually not a requirement to hear signals which are only just above the noise level.

In many receivers it is necessary to ensure that the receiver stays on the correct frequency. Any drift in the set may mean that the signal may not remain on the linear portion of the demodulator curve. Some sets incorporate a circuit called an automatic frequency control or AFC. This enables the set to remain on tune, even when the local oscillator is prone to drifting. The

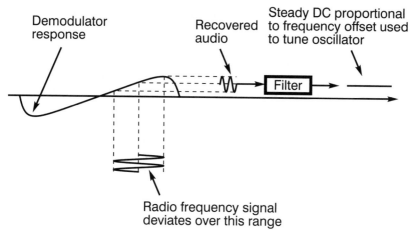

Figure 5.40 *Operation of an AFC*

circuit operates by taking the output from the demodulator and removing the audio using a simple filter. The remaining voltage is then an indication of where the carrier is on the demodulator curve. Ideally it should be in the centre. The voltage can be applied to the local oscillator in the set as a correction signal to pull its frequency slightly so that the signal remains in the centre of the pass-band.

One of the advantages of FM is its resilience to noise. This is one of the main reasons why it is used for high quality audio broadcasts. However, when no signal is present, a high noise level is present at the output of the receiver. If a low level FM signal is introduced and its level slowly increased it will be found that the noise level reduces. From this the quieting level can be deduced. It is the reduction in noise level expressed in decibels when a signal of a given strength is introduced to the input of the set. Typically a broadcast tuner may give a quieting level of 30 dB for an input level of just over a microvolt.

Another effect which is often associated with FM is called the capture effect. This can be demonstrated when two signals are present on the same frequency. In cases of this nature, only the stronger signal will be heard at the output, unlike AM where a mixture of the two signals is heard, along with a heterodyne if there is a frequency difference.

The capture effect is often defined in receiver specifications as the capture ratio. It is defined as the ratio between the wanted and unwanted signal to give a certain level of the unwanted signal. Normally a reduction of the unwanted signal of 30 dB is used. To give an example the capture ratio may be 2 dB for a typical tuner to give a reduction of 30 dB in the unwanted signal. In other words if the wanted signal is only 2 dB stronger than the unwanted one, the audio level of the unwanted one will be suppressed by 30 dB.

A number of circuits can be used to demodulate FM. To illustrate the methods which are used, circuits using discrete components are given, although in reality most radios these days use integrated circuits. Two circuits which were widely used were called the ratio detector or demodulator and the Foster-Seeley demodulator.

The circuit of the ratio detector is shown in Figure 5.41. It is characterized by the special transformer and the diodes which are effectively in series with one another. When a steady carrier is applied to the circuit the diodes act to produce a steady voltage across the resistors R_1 and R_2, and the capacitor C_3 charges up as a result.

The transformer enables the circuit to detect changes in the frequency of the incoming signal. It has three windings. The primary and secondary act in the normal way to produce a signal at the output. The third winding is untuned and the coupling between the primary and the third winding is very tight, and this means that the phasing between signals in these two windings is the same.

The primary and secondary windings are tuned and lightly coupled. This means that there is a phase difference of 90° between the signals in these windings at the centre frequency. If the signal moves away from the centre frequency the phase difference will change. In turn the phase difference between the secondary and third windings also varies. When this occurs the voltage will subtract from one side of the secondary and add to the other causing an imbalance across the resistors R_1 and R_2. As a result this causes a current to flow in the third winding and the modulation to appear at the output.

The capacitors C_1 and C_2 filter any remaining RF signal which may appear across the resistors. The capacitor C_4 and R_3 also act as filters ensuring no RF reaches the audio section of the receiver.

The Foster-Seeley detector is very similar to the ratio detector in many ways, but there are a few fundamental differences. Although the circuit topology looks very similar, having a transformer and a pair of diodes, there is no third winding. Instead this is replaced with a choke.

While there are a number of differences the circuit again operates on the phase difference between signals. To obtain the different phased signals a connection is made to the primary side of the transformer using a capacitor, and this is taken to the centre tap of the transformer. This gives a signal which is 90° out of phase.

When an unmodulated carrier is present at the centre frequency, both diodes conduct, producing voltages across their respective load resistors which are equal and opposite. These two voltages cancel one another out at the output so that no voltage is present. As the carrier moves off to one side of the centre frequency the balance condition is destroyed, and one diode will

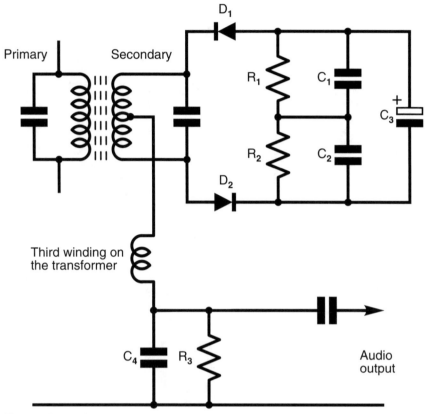

Figure 5.41 *Circuit of the ratio detector*

conduct more than the other. This results in the voltage across one of the resistors being larger than the other, and a resulting voltage at the output corresponding to the modulation on the incoming signal.

The choke is required in the circuit to ensure that no RF signals appear at the output. The capacitors C_1 and C_2 provide a similar filtering function.

One of the major problems with both of these circuits is that they require coils and chokes. These are expensive to produce these days, often costing more than an integrated circuit. As a result other methods of demodulating FM are often used.

Phase locked loops are widely used for demodulating FM. Apart from not using a coil they also give a very linear voltage to frequency conversion. As a result of this they are often used in hi-fi tuners.

The way in which they operate is very simple. The circuit is set up to operate as shown in Figure 5.43. The FM signal from the IF stages of the set

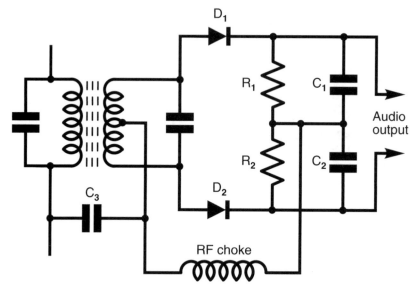

Figure 5.42 *Circuit of the Foster-Seeley detector*

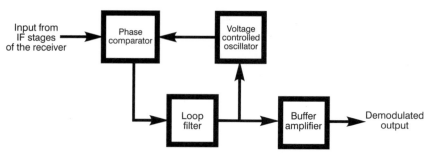

Figure 5.43 *A phase locked loop FM demodulator*

is connected to one of the phase detector inputs as shown, and the output from the VCO is connected to the other.

With no modulation applied and the carrier in the centre position of the pass-band the voltage on the tune line to the VCO is set to the mid position. However, if the carrier deviates in frequency, the loop will try to keep the loop in lock. For this to happen the VCO frequency must follow the incoming signal, and for this to occur the tune line voltage must vary. Monitoring the tune line shows that the variations in voltage correspond to the modulation applied to the signal. By amplifying the variations in voltage on the tune line it is possible to generate the demodulated signal.

It is found that the linearity of this type of detector is governed by the voltage to frequency characteristic of the VCO. As it normally only swings over a small portion of its bandwidth, and the characteristic can be made relatively linear, the distortion levels from phase locked loop demodulators are normally very low. Many tuners have specifications of fractions of a percent of distortion.

Digital signal processing

Microprocessors are finding uses in an increasing number of applications. It is therefore hardly surprising to find that they are being used in a variety of areas which normally used only analogue electronics. Functions including filtering, demodulation and the like can all be performed digitally by specialized digital signal processors. As a result some of the higher priced receivers are beginning to use these techniques, and in the coming years this process will become far more widespread.

The process is based upon the fact that it is possible to build up a representation of the signal in a digital form by sampling the voltage level at regular time intervals and converting the voltage level at that instant into a digital number proportional to the voltage as shown in Figure 5.45. This

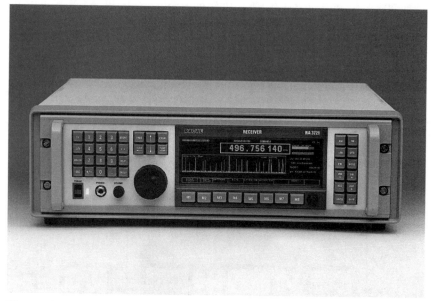

Figure 5.44 *A high performance wide-band communications receiver which uses digital signal processing (Courtesy of Racal Radio)*

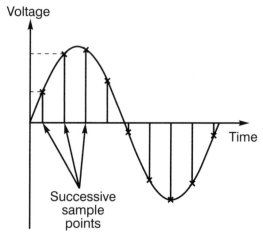

Figure 5.45 *Sampling a waveform*

process is performed by a circuit called an analogue to digital converter, A to D converter or ADC.

Once in a digital format the processor performs complicated mathematical routines upon the representation of the signal. However, to use the signal it then needs to be converted back into an analogue form where it can be amplified and passed into a loudspeaker or headphones as shown in Figure 5.46. The circuit which performs this function is, not surprisingly, called a digital to analogue converter, D to A converter or DAC.

The advantage of digital signal processing is that once the signals are converted into a digital format they can be manipulated mathematically. This gives the advantage that all the signals can be treated far more exactly, and this enables better filtering, demodulation and general manipulation of the signal. Unfortunately it does not mean that filters can be made with infinitely steep sides because there are mathematical limitations to what can be accomplished.

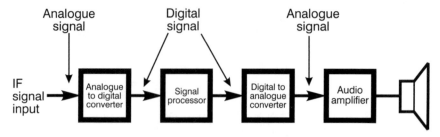

Figure 5.46 *Block diagram of a digital signal processor*

Sensitivity and noise

One of the most important aspects of any radio receiver is its sensitivity, or ability to pick up weak signals. Most receivers today are very sensitive and able to pick up signals which are less than a microvolt at the input to the set. Unfortunately a set cannot be made more sensitive by simply adding more stages of amplification as a number of other aspects of receiver design quickly become obvious and limit the sensitivity. The first is that if there is too much gain then some of the stages in the set become overloaded leading to other problems which will be described later. The other problem is that of noise and this is the major limiting factor. There is a certain amount of noise which is present at the input to the receiver, and each stage within the set adds a little more.

Noise consists of random electrical impulses and is present at all frequencies, although the levels and sources vary with frequency. In many cases it is 'white noise' which is heard as a background hiss and it is present at all frequencies. Once it is present it cannot be removed or cancelled out. It can only be reduced by limiting the bandwidth of the set, and if the bandwidth is reduced too much then the wanted signal may be impaired.

Other types of noise may appear as bangs or pops, and this type of noise is heard particularly on the lower frequency bands.

Noise comes from two main areas. It may be picked up by the aerial. In turn this can come from a variety of sources. It may be man-made. The variety of electrical and electronic equipment in use today generates noise which can be picked up over the whole of the radio spectrum. Electric motors, vehicle ignition systems, televisions, computers and even fluorescent lights all produce energy which can be picked up.

Naturally occurring noise can also be picked up. Again there are a variety of sources for this. Static is one major type of noise. This comes from discharges in electrical storms. In view of the colossal amounts of energy involved in these discharges it is hardly surprising they generate large amounts of radio frequency energy which can be heard over large areas. Cosmic or galactic noise comes from many places in outer space, from our own sun, to the distant galaxies. Surprisingly this noise can be present at significant levels, despite the enormous distances involved.

The level of different types of noise varies with frequency, and this means that the design of radios is dependent to a certain degree upon this. Some forms of man-made noise extend to relatively high frequencies. Car ignition noise can extend to frequencies in excess of 400 MHz and noise from fluorescent lights can also be picked up at frequencies well above 1 GHz.

Naturally occurring forms of noise fall with increasing frequency. Atmospheric noise predominates at low frequencies, but above about 10 MHz galactic noise is the main constituent. Even this falls with increasing frequency, leaving that generated within the receiver as the main source of noise at frequencies above about 50 MHz.

Receiver noise performance is fashioned to a large degree by the levels of noise which occur. In view of the high levels of noise entering a receiver from the antenna on frequencies below 30 MHz there is little point in designing very low noise receivers. At frequencies above about 50 MHz to 100 MHz received noise levels fall to a level where the noise generated within the receiver predominates, and accordingly the noise performance of the set becomes far more important.

Another form of noise is called thermal noise. This is caused by the movement of the electrons in a conductor. Even when there is no potential difference to cause current to flow, noise is present. At any temperature above absolute zero electrons are moving about randomly in the conductor, and this causes random voltages or noise to be generated. It is not surprising that the level of noise is proportional to the temperature. As the temperature rises so the movement of the electrons increases. It is also found that the level of noise is proportional to the bandwidth being received. The wider the bandwidth, the greater number of noise frequencies which can be received, and hence the greater the level of noise. Finally the noise is proportional to the resistance of the conductor. In fact the level of noise can be calculated from the formula given below.

$$E = \sqrt{4kTBR}$$

where E = EMF in volts
k = Boltzmann's constant which is 1.37×10^{-23} joules per degree Kelvin
T = Absolute temperature in degrees Kelvin
B = Bandwidth in hertz
R = Resistance in ohms

Unfortunately there is no way of reducing the noise generated by a resistor, except by reducing the resistance, the bandwidth or its temperature. It is for this reason that some very low noise amplifiers are cooled.

Signal to noise ratio

There is a variety of methods used to measure and define the sensitivity of a radio receiver. As the noise performance is the limiting factor, the sensitivity is specified in terms of its noise performance. The most obvious method of achieving this is to measure the difference between the wanted signal and the noise level under specified conditions, as shown in Figure 5.47.

The difference between the wanted signal and the noise is expressed as a ratio in decibels. This is often termed the signal to noise or S/N ratio. The signal level also needs to be mentioned as this has a direct bearing on the

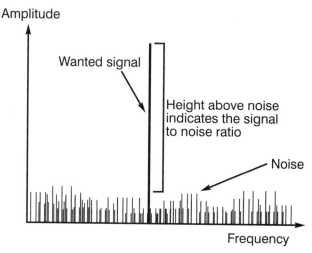

Amplitude

Wanted signal

Height above noise
indicates the signal
to noise ratio

Noise

Frequency

Figure 5.47 *Signal to noise ratio*

figures obtained. The bandwidth of the receiver also has a bearing on the figures which are obtained because the noise level is proportional to the bandwidth being used.

When specifying the performance for AM the level of modulation also needs to be included in the specification. This is because the audio output from the receiver is measured to give the signal level and this will be dependent upon the modulation level. The usual value for modulation is 30%.

A typical specification for a good short wave band communications receiver may be in the region of 0.5 microvolts for a 10 dB S/N in a 3 kHz bandwidth for SSB and Morse. For AM reception it may be in the region of 1.5 microvolts for a 10 dB S/N in a 6 kHz bandwidth with a 30% modulation level. Note that a wider bandwidth is required for AM reception and this means that the noise level is higher and sensitivity is less.

In some instances a signal plus noise-to-noise (S + N)/N ratio is specified. The reason for this arises out of the way the measurement is made. A signal generator is connected to the input of the receiver, and at the audio output a meter is used to measure the audio level. With the signal generator turned off the level of the noise is noted at the output. Then the signal generator output is turned on and its level is adjusted so that the output from the receiver is 10 dB higher than the noise level.

When the signal generator is turned on the audio level meter is reading the level of the output signal plus any background noise, i.e. (S + N)/N. Note that during the measurements for this test when the signal generator output is turned off the impedance seen by the radio receiver must be 50 ohms. This is normally true because the output attenuators and other matching circuits on the signal generator ensure the output is maintained at this value.

Occasionally the signal generator specification will mention that the signal voltage is given as a potential difference (PD) or the electromotive force (EMF). This is very important because there is a factor of two difference between them. The EMF is the open circuit voltage whereas the PD is the voltage under load as shown in Figure 5.48. If the load impedance is the same as the source impedance the EMF will be twice the PD at the load. In theory the EMF is a more correct way of specifying the voltage because the way in which the signal generator works assumes that it has been loaded with 50 ohms for the value to be correct. However, as the values with the PD look more impressive, it is this value which is normally stated. Even when EMF or PD is not specifically mentioned it should be assumed that PD is implied.

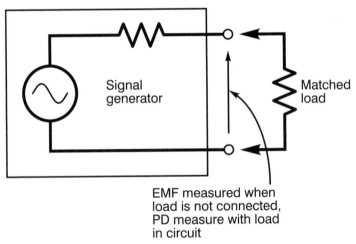

EMF measured when load is not connected, PD measure with load in circuit

Figure 5.48 *EMF and PD in a signal generator*

SINAD

Sensitivity may also be specified in terms of SINAD measurements. This is very similar to a signal plus noise to noise measurement. The test method involves applying a signal modulated by a single tone. The audio from the receiver and the audio tone is notched out as required. In this way a measurement of signal plus noise plus distortion to noise plus distortion is obtained. The SINAD value is expressed in decibels.

Using this system the sensitivity of a set is normally quoted as a given number of microvolts to give a certain value of SINAD. A figure of 12 dB

SINAD is normally used as this represents a 25% distortion and noise factor.

SINAD measurements are most commonly used for FM sets. However, they can be used for other modes as well. For AM it is simply a matter of changing the type of modulation. For SSB it is necessary to ensure that the receiver is tuned to exactly the correct frequency so that the audio tone can be notched out. Once this has been done the measurement can be made in the normal way.

Noise factor and noise figure

Apart from signal to noise ratio and SINAD specifications, the noise performance can be specified in terms of a noise figure. This measurement is more versatile than the signal to noise ratio because it can be used to determine the performance of a piece of equipment whether it is a complete system, a receiver, or a smaller item such as a preamplifier. Essentially this measurement gives an indication of the level of noise which each item introduces.

The idea of noise figure and noise factor is built around the fact that any system has a certain amount of noise below which it is not possible to go. This thermal noise is dependent upon the resistance of the system. A 50 ohm resistor generates a certain level of thermal noise. As an antenna looks like a resistor to the input of the receiver, this too generates noise of its own.

As a result of this thermal noise, any signal being picked up by the antenna has a certain signal to noise ratio associated with it. If the signal is very strong the signal to noise ratio will be good, but if it is weak it will be poor. When the signal is passed through an amplifier, receiver or any other piece of electronic equipment the signal to noise ratio will be degraded because the circuit will introduce additional noise.

The noise factor is determined by taking the signal to noise ratio at the input and dividing it by the signal to noise ratio at the output. For these calculations the signal to noise ratio must be given as a ratio and not

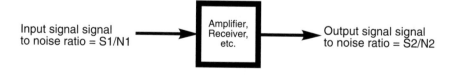

$$\text{Noise figure} = 10 \log_{10} \frac{S1/N1}{S2/N2}$$

Figure 5.49 *Noise figure of a system*

expressed in decibels. As the circuit will always degrade the signal the noise factor is always greater than one. The noise figure is obtained by converting the noise factor into decibels, as shown in Figure 5.49.

To give an example of this a signal may have a signal to noise ratio of 6:1 at the aerial, and this may be degraded to 4:1 after passing it through an amplifier. For this amplifier the noise factor would be 6/4 and the noise figure would be $10 \log_{10} 1.5$ or $1.76\,dB$.

Strong signal and overload

While the sensitivity of a receiver is very important it is not the only aspect which designers have to consider. Equally important is the way in which the radio is able to handle strong signals. A receiver may need to receive weak signals which are close to very strong ones. It is quite possible that a set may need to receive signals which vary in strength by up to $100\,dB$. This enormous variation in strength tests any design to its limits, particularly the front end stages.

Under normal conditions the amplifiers in a set must remain linear. In other words the output is directly proportional to the input. However, when strong signals are received there comes a point where the output of the amplifier starts to overload and cannot give out the required level. At this point the amplifier is said to be in compression.

Compression itself is not a problem as the absolute values of a signal are rarely important and in any case the AGC acts on the signal level to alter the gain of the set. However, it is the effects associated with compression and non-linearity which cause the problems. Intermodulation, blocking, cross-modulation are three problems associated with overload, and these can noticeably reduce the performance of a receiver in some circumstances.

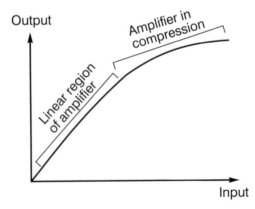

Figure 5.50 *Characteristic of a typical amplifier*

Intermodulation distortion

One of the main problems associated with overload is intermodulation distortion. When this occurs, new signals are generated within the receiver, sometimes giving the perception that it is receiving a lot of signals and it is very sensitive.

Two effects occur to give rise to intermodulation distortion when an amplifier becomes overloaded. One is that harmonics of signals being received are generated, and the second is that the amplifier, being non-linear, acts as a mixer. On their own these effects are unlikely to degrade the performance very much because the RF tuning would remove signals which would pass through the rest of the set. If a signal at frequency f is generating harmonics, these will fall at frequencies of $2f$, $3f$, $4f$, and so on. For even the lowest harmonic to enter the set, the fundamental at half the frequency must enter the front end. The RF tuning will reduce this to a low level and it is unlikely to cause a problem for most applications. Similarly if two signals are to mix together to form a signal which is within the receiver pass-band they will be outside the acceptance range of the front end and their effects are not noticed under normal conditions. Being just a mix product between two signals this is called a second order effect.

The major problem occurs when harmonic generation and mixing occur together. It is possible for the harmonic of one signal to mix with the harmonic or fundamental of another to give a third signal which is within the pass-band of the receiver, i.e. $2f_1 - f_2$. It is only the difference products which cause a problem, as the sum products fall well away from the received frequencies. To give an example, signals on 100.00 and 100.1 MHz may enter

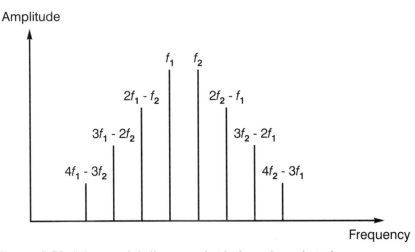

Figure 5.51 *Intermodulation products from two signals*

the front end of a receiver. The harmonic of the first will be at 200.00 MHz and this can mix with the second (200 – 100.1) to give a signal at 99.99 MHz. All of these signals are well within the pass-band of the front end and the unwanted signal at 99.9 MHz will appear as a real signal. This is a third order effect. Other higher order products can be calculated and it can be seen that a comb like that shown in Figure 5.51 is produced. Note that only the odd order effects cause problems.

Third order intercept

To give an indication of the performance of an amplifier a figure known as the third order intercept point is often quoted. It is found that the levels of distortion in a circuit are very small under normal operating conditions, but rise very rapidly as the level of the input signal increases. For every 1 dB increase in signal level a third order signal will rise by 3 dB and a fifth order signal by 5 dB and so on.

Normally the amplifier would run into saturation well before the third order effect became comparable with the wanted signal. However, it is possible to plot the curves the levels of the two signals would take. At a certain point they intersect, as shown in Figure 5.52. This is known as the third order intercept point, and it gives an indication about its strong signal handling capacity. Typically a good short wave or communications receiver may have an intercept point of around 25 dBm (i.e. a signal which is 25 dB above a milliwatt).

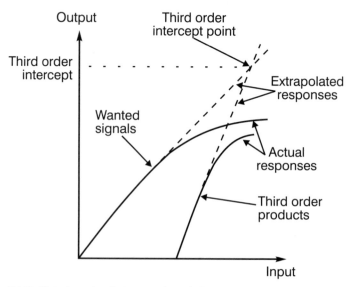

Figure 5.52 *Third order intercept point*

Blocking

Often a very strong signal which is off-channel can reduce the sensitivity of a receiver. This happens because the RF amplifier stage is being driven into compression. When this occurs it has the effect of only allowing the main signal through the amplifier, and all the others are reduced in strength. When blocking is quoted it is usually given as the level of signal 20 kHz away from the received channel which gives a 3 dB reduction in the wanted signal. Usually a good set will be able to withstand levels of several milliwatts before this happens.

Cross-modulation

Another effect caused by overloading is called cross-modulation. When this occurs the modulation or amplitude variations on a strong signal are superimposed on signals close by. It is particularly noticeable on AM signals where the modulation of a strong signal is superimposed onto weaker signals either side.

The problem is a third order effect, and often occurs as a result of poor mixer performance. However, any set with a high third order intercept point should give good cross-modulation performance as well.

Dynamic range

The dynamic range of the most important parameters of its specification as it outlines the range over which it can operate. In some instances it may be necessary for a set to be able to accommodate signals with enormous differences in strength, especially if a good antenna is used.

There are a number of ways in which the dynamic range may be determined. Essentially it is the difference between the weakest signal which it can hear and the strongest one it can tolerate without any noticeable degradation in performance. As these two end points can be specified in a number of different ways they must be included in the dynamic range specification.

The low end of the range is governed by the sensitivity of the set. Usually the minimum discernible signal (MDS) is used. This is the weakest signal that the set can hear, and it is usually taken as a signal equal to the noise produced by the set. Often the signal may be around –135 dBm for a 3 kHz bandwidth.

At the other end of the range there are two main limiting factors. One is the onset of blocking, and this may be at the point where sensitivity is reduced by either 1 dB or 3 dB. The other is the generation of intermodulation products. Here the point which is often taken is the level of input signals which generate signals which could mask a signal equal to the minimum discernible signal, i.e. the noise floor.

Typically sets have an intermodulation limited dynamic range of between 80 and 90 dB whereas the blocking limited dynamic range may be around 110 dB or more.

The dynamic range is usually limited by the performance of the front end stages of the set. Not only is the noise performance of these stages critical, but so too is the strong signal handling performance.

6 Transmitters

Today there is an enormous variety of different types of transmitter used to generate all the signals which are to be heard on the radio bands. Their purpose is to generate the basic signal or carrier, and then superimpose the modulation onto it in the correct format. Once this is done the signal is amplified to the correct level and filtered to remove any spurious products which are outside the required band. At this stage the output stages may also include matching circuitry to ensure that there is an accurate impedance match between the load and the transmitter. In this way the maximum power transfer takes place.

Transmitter building blocks

Like all electronics the complexity of transmitters is increasing, and their flexibility is improving to enable them to fulfil the increasing requirements placed on them. From low power UHF transmitters in cellular phones to the high power broadcast stations, they all use a number of the same basic building blocks to achieve the output of the required signal.

Oscillator

The heart of any transmitter is its master oscillator. This generates the carrier onto which the modulation is superimposed. Many transmitters utilize a number of oscillators to mix the signal to its final required frequency. However, this processing normally takes place after the modulation has been applied.

The requirements of the oscillator wherever it is used in the equipment are that it should be stable and not drift. If it does drift then the output signal will change. In some instances drift on an oscillator may cause the characteristics of the signal to change, particularly in the case of an SSB transmitter.

The oscillators used in transmitters take many forms. Those not requiring to change frequency often use crystals to ensure high degrees of stability. In some cases where high degrees of accuracy and stability are required, crystal ovens may be employed.

For signals which need to change in frequency a variable frequency oscillator (VFO) is required. The simplest form of VFO uses an inductor and variable capacitor. An example of a suitable oscillator is shown in Chapter 5, Figure 5.17. The output from this circuit must be buffered using a further amplifier. This will reduce the effects of any changes in load which may occur and cause frequency shifts. This is particularly true for the more simple circuits like the Morse transmitter described later where the difference between the key up and key down states may cause a frequency change. Buffer amplifiers generally have a high input impedance to ensure that the oscillator circuit is not loaded to any degree.

More sophisticated oscillator systems may use a crystal mixer system similar to that employed in a superhet radio. Here a VFO running at a relatively low frequency is mixed with a crystal oscillator running at a higher frequency. If wide-band operation is required the crystal oscillator can be

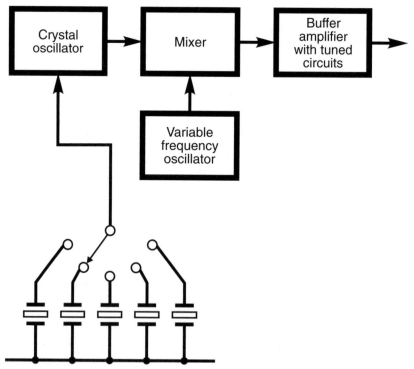

Figure 6.1 *Block diagram of a crystal mixer VFO*

switched to increase the frequency coverage. The inherent stability of the crystal oscillator enables the whole oscillator system to remain stable despite the fact that part of the circuit is switched. Also by keeping the variable oscillator running at a relatively low frequency the stability is maintained. However, the output of a VFO which uses a mixer must be tuned to prevent unwanted or spurious signals from entering the later stages of the transmitter and ultimately being radiated.

Today most transmitters which need variable frequency capability use a frequency synthesizer. This gives much greater stability while still being able to operate over a wide range of frequencies. Synthesizers for use in transmitters are exactly the same as those employed in receivers. In transceivers the same one is used for both transmit and receive functions. Synthesizer phase noise is also important in transmitters. In the same way that receivers suffered from reciprocal mixing as a result of phase noise generated by a synthesizer, a poorly designed synthesizer will generate wide-band noise extending either side of the carrier. This will be transmitted and may cause interference to other users nearby if the signal is strong enough. As a result care must be taken in the design to ensure that this does not occur.

Amplifiers

The output from the oscillator on its own is not enough to enable communications to be established. The power level may only be a milliwattt or so and an amplifier or series of amplifiers is required to bring the signal up to the required level. Cellular phones often have an output power of up to 3 or 4 watts. At the other extreme high power broadcast stations may have outputs in the region of 100 kW and some may even exceed a megawatt.

In small transmitters including hand-held transceivers, mobile phones and the like the amplifiers are all semiconductor based. Even many high power transmitters still use semiconductors, but those which deliver many kilowatts use thermionic technology (valves) as they are still the most convenient way of generating these high levels of RF power.

In many ways an RF amplifier is like any other form of amplifier. A signal enters at the input and an amplified or larger version appears at the output.

In most cases it is a requirement that an amplifier is as linear as possible. The transfer characteristic of the amplifier should be totally linear so that the output is a faithful representation of the input. In some transmitter applications this may not be the case. Good linearity is accomplished at the expense of poor power efficiency. In some cases efficiency is very important. For portable applications it is necessary to ensure that as much of the DC power entering the amplifier is converted into radio frequency power that can be transmitted. In this way the batteries can be made to last as long as possible. At the other end of the spectrum broadcast stations with transmitters that consume many hundreds of kilowatts find that the cost of the electricity is a major consideration. By increasing the efficiency of the amplifiers, significant cost savings can be made.

The efficiency of the amplifier is simply the proportion of the DC input power supplied to the amplifier which is converted into radio frequency power to be transmitted. This can be expressed:

$$\eta = \frac{\text{Power output}}{\text{Power input}} \times 100\%$$

There are a number of factors which affect the efficiency of an amplifier. One of these is the mode of its operation. Those which operate in a totally linear mode are less efficient than those which run in a less linear mode. As a result amplifiers are classified in terms of their mode of operation. Normal

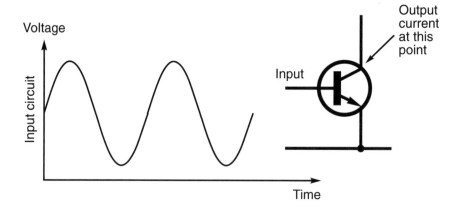

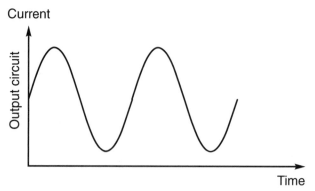

Figure 6.2 *Class A operation*

linear amplifiers are said to operate in class A. They accomplish a high degree of linearity at the expense of poor efficiency. The reason for this is that in order for the amplifier to maintain a linear relationship between the input and the output it must conduct current for the whole of the input signal cycle, as shown in Figure 6.2. The maximum possible efficiency for a class A amplifier is 50%, with values between 25 and 45% being more normal.

Efficiency can be improved by biasing the amplifier to conduct for only part of the cycle. For class B operation the amplifier conducts over only half the cycle, as shown in Figure 6.3. If an amplifier of this type is operated into a resistive load it effectively rectifies the signal, passing only one half of the cycle. Sometimes this type of circuit may even be used as a simple detector.

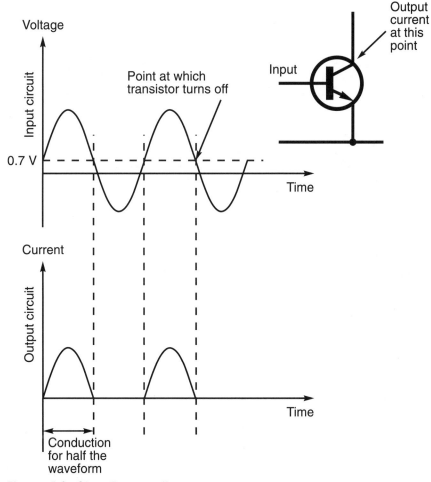

Figure 6.3 *Class B operation*

Another application is to use two class B amplifiers operating over opposite halves of the cycle. By summing the outputs both halves of the cycle are present, and a linear representation of the signal is produced. This type of operation is called push–pull, because each half of the amplifier operates on a different half of the cycle.

For RF applications a single ended class B amplifier may be used in conjunction with a tuned circuit. In operation the tuned circuit 'rings' when the half wave signal is applied, and this supplies the missing half of the cycle. Amplifiers of this type typically have an efficiency of around 60%.

To improve the efficiency still further, the amplifier may be operated in class C. In this mode of operation the amplifier is biased to allow short

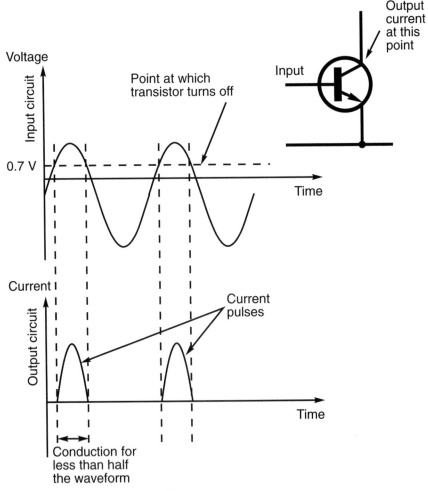

Figure 6.4 *Class C operation*

current pulses of less than half a cycle to flow, corresponding to the peaks in the input waveform. In view of their mode of operation class C amplifiers are very non-linear and the output is rich in harmonics. This means that a filter is required to remove them before being passed to an antenna. The filter is tuned to the frequency of operation and only allows through the required frequency. In this way the required sine wave is extracted.

Class C amplifiers perform very well as frequency multipliers. Their output is particularly rich in odd harmonics, extending to many times the fundamental frequency. However, it is normal to only select the third or at the highest the fifth harmonic. This is achieved using the filter in the amplifier output circuit. If higher order harmonics are selected it is not normally possible to achieve sufficient rejection of the unwanted harmonics, and this results in unwanted signals being transmitted.

One very important aspect of a transmitter power amplifier is the amount of power which it can deliver to the antenna system. Radio frequency power is measured in exactly the same way as low frequency AC power. It is simply the voltage times the current, assuming they are in phase with one another. Assuming that the antenna load is resistive then they will be in phase. Very high power amplifiers like those used as the final stage for a broadcast transmitter may be capable of delivering 100 kilowatts or more to the antenna, whereas those used for local broadcast stations may only deliver 100 watts or so. Other types of transmitter can deliver a variety of power levels dependent upon their use. Cellular phones deliver a few watts, CB transmitters are generally limited to less than 5 watts. Amateur radio ones are limited to a maximum of a kilowatt or so dependent upon the restrictions of the individual country, although many operators enjoy operating with powers of a watt or sometimes even less.

Power levels are often measured directly in watts. However, they are often specified in decibels relative to a watt or a millliwatt. Given the units dBm (decibels relative to a milliwatt) and dBW (decibels relative to a watt), they are being used increasingly, especially in professional circles. Using this system it is very easy to incorporate any changes in level brought about by the use of amplifiers, attenuators, losses in feeders, etc. All of these are generally expressed in decibels so it is simply a matter of adding or subtracting the relevant figure from the power level, e.g. a power of 10 dBW becomes only 6 dBW at the end of a feeder with a 4 dB loss. For transmitters, the power levels are generally expressed in dBW, whereas levels in dBm are generally used for low power circuits. For example, a diode ring mixer may require a local oscillator input of 7 dBm.

It is important to ensure that an amplifier operates under the correct conditions. When there is a poor impedance match between the amplifier and its load it will operate under conditions which are not ideal. Amplifiers are designed to operate into a given load, normally 50 ohms for which feeders are available. The antenna attached to this may not present a good load. When

Table 6.1   dBm – dBW – watts conversion

dBm	dBW	Watts	Terminology
+100	+70	10 000 000	10 megawatts
+90	+60	1 000 000	1 megawatt
+80	+50	100 000	100 kilowatts
+70	+40	10 000	10 kilowatts
+60	+30	1 000	1 kilowatt
+50	+20	100	100 watts
+40	+10	10	10 watts
+30	0	1	1 watt
+20	−10	0.1	100 milliwatts
+10	−20	0.01	10 milliwatts
0	−30	0.001	1 milliwatt
−10	−40	0.0001	100 microwatts
−20	−50	0.00001	10 microwatts
−30	−60	0.000001	1 microwatt
−40	−70	0.0000001	100 nanowatts
−50	−80	0.00000001	10 nanowatts
−60	−90	0.000000001	1 nanowatt

this happens power is reflected back to the amplifier and high levels of VSWR may be present (see Chapter 7). Under these conditions damage may occur to the amplifier because high levels of current or voltage may be present. If a high voltage occurs, this may cause the maximum voltage for the device to be exceeded with resultant breakdown of the device. Alternatively high current levels can cause its current ratings to be exceeded resulting in the burn-out of the device. In view of this many amplifiers incorporate protection circuitry which reduces the power level as the VSWR increases. Other protection systems may remove the power from the amplifier if a high VSWR is seen. Often low power amplifiers have sufficient margin to withstand most levels of VSWR. If protection circuitry is required, then it must be very fast acting. Output transistors can be destroyed very quickly, especially if voltage breakdown occurs.

Mixers

The basic mixing process and the circuits used have been described in Chapter 5. Not only are they used widely for receiver applications, but they are also widespread in transmitter circuitry as well. Here they are also used in frequency conversion applications to change a signal from one frequency to another. In addition to this they are used in the modulation process, especially for single sideband generation as described later. In the generation of these signals balanced mixers which only have the mix products at the output, with the original input signals suppressed, are particularly useful, as will be seen later.

Filters and matching networks

The networks used to filter and match transmitters are very important. Being based around circuits using inductors and capacitors these networks normally perform the functions of filtering and matching together. The operation of these filters is described in Chapter 3 (Capacitors, inductors and filters).

Often within the transmitter itself there will be a number of filters, especially after mixers, and these ensure that only the required signals are passed onto the next stage. At the output of the transmitter these networks are particularly important. In the first instance the matching circuitry ensures that the impedance match between the power amplifier and the feeder is optimum. This enables the maximum amount of power to be transferred from the amplifier to the feeder and hence radiated. The filter ensures that the radiated signal does not contain any undue levels of unwanted spurious signals. As most amplifiers generate certain levels of harmonics, most power amplifiers need good low pass filtering.

Simple Morse transmitter

An effective Morse transmitter can be made out of very few components. As Morse signals can be copied at lower strengths than other modes, powers can be relatively low. Coupled to this the circuitry can be kept simple if few facilities are needed. Indeed it is possible to make a Morse transmitter with only two or three transistors and a handful of other components. A number of amateur radio enthusiasts use transmitters like these and can make contact with places all over the world using very modest powers and simple home-built equipment.

Figure 6.6 shows the block diagram of the very basic Morse transmitter. All that is required is an oscillator to generate the signal, a power amplifier to increase its level, a method of switching the signal on and off, and finally the matching and filtering circuitry at the output.

In the very simplest circuits the oscillator may be crystal controlled to ensure sufficient stability, although this does limit the flexibility of the set. Often a VFO is employed to enable the set to operate over a band of frequencies. This could be an LC tuned oscillator or a more complicated unit based around a synthesizer. However, for a transmitter of this simplicity it is unlikely that a sophisticated synthesized oscillator would be used.

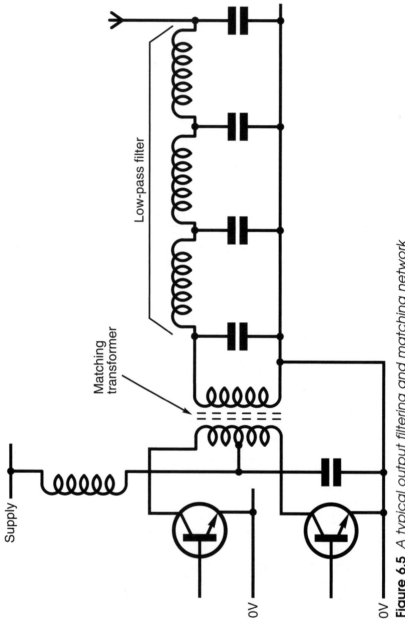

Figure 6.5 *A typical output filtering and matching network*

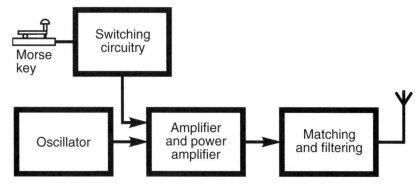

Figure 6.6 *Block diagram of a simple Morse transmitter*

The power amplifier is used to increase the level of the signal. In this type of transmitter where no analogue modulation is applied to the carrier, the PA will be operating in class C to give the maximum efficiency.

The final stage is the filtering and matching circuitry. For a low power transmitter it is often possible for the output device to give a good match to 50 ohms with little or no impedance transformation. As a result only filtering may be required. For a transmitter of this nature this will consist of a low pass filter. This will remove the spurious signals which will be harmonics of the required signal. Typically this may be a five pole filter with a cut-off just above the operational frequency.

Amplitude modulation transmitter

A huge number of signals in the radio spectrum are modulated to carry audio. The simplest way of achieving this is to modulate the amplitude of the carrier in line with the audio signal. While this method is not as efficient as many other methods, it is still widely used for broadcast stations in the long, medium, and short wave bands as well as being used by aircraft at VHF.

A basic AM transmitter requires the carrier to be generated and this is modified by the modulating signal. Figure 6.7 shows a basic AM transmitter. In this circuit a system called high level modulation is used; in other words the modulation is applied to the carrier in the final amplifier.

The carrier is generated by the oscillator circuit. This may be a variable frequency oscillator, or possibly a crystal oscillator. Today frequency synthesizers are often likely to be used, for convenience and flexibility.

Once the basic signal has been generated it is buffered and amplified. This brings it to the correct level for the signal to drive the final power amplifier. Both the drive amplifiers and the final amplifier will operate in class C to maintain the optimum efficiency.

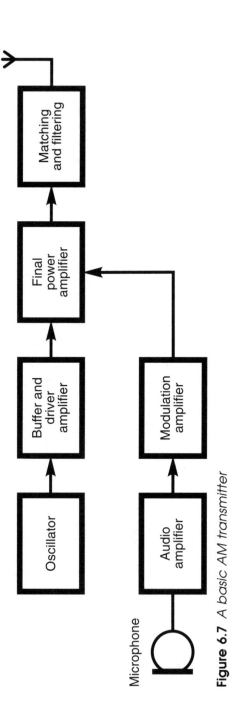

Microphone

Figure 6.7 *A basic AM transmitter*

The audio signal from the microphone or other source enters the transmitter and is amplified. These stages may also limit or process the signals to improve their intelligibility, or reduce the bandwidth occupied. For most communications purposes frequencies above 3 kHz and below 300 Hz are attenuated. For broadcast purposes top limits above this are used dependent upon the channel spacing in use. On the short wave bands this is 5 kHz and on the medium wave band it is 9 kHz in Europe and 10 kHz in North America.

The audio signal is further amplified by a driver amplifier if required and applied to the modulation amplifier. This is a high power audio amplifier which must be capable of developing an audio power equal to half the input power requirements for the final RF power amplifier if 100% modulation is to be achieved. In other words if the RF amplifier consumes 100 watts, then the audio amplifier must be able to deliver 50 watts. The power audio signal is then used to modulate the supply to the final RF power amplifier and in this way modulate the RF signal.

If any further increase in the power level of the transmitter is required after the modulation has been applied, the RF amplifier must run in class A to prevent distortion of the modulation. These amplifiers are often called linear amplifiers or 'linears' in view of their mode of operation.

Figure 6.8 *A high power short wave AM broadcast transmitter (Courtesy of the BBC)*

The final stage of the transmitter, like that of the Morse transmitter is to pass the signal through the matching and filtering circuits. Again these are required to reduce the levels of spurious signals to acceptable limits.

Single sideband transmitter

There are a number of different methods in which a single sideband signal can be generated. The first and most commonly used employs a filter to remove the unwanted sideband. A second method uses phasing techniques to eliminate the unwanted elements in the signal.

The filter method is virtually a reverse of the signal path followed in a typical receiver. This is the most common method used as it is capable of giving excellent results. It is also very convenient for use in transmitter-receivers (transceivers) because many of the circuits can be used in both the transmit and receive paths, saving on production costs.

The outline of an SSB transmitter using the filter method is given in Figure 6.9. The first block in the transmitter is the carrier oscillator. This is normally crystal controlled because its frequency stability is critical to maintain the correct frequency relative to the filters used later.

Audio signals from the microphone are amplified and processed as necessary and applied with the carrier to a balanced mixer circuit. The advantage of using this type of is that the input signals are suppressed at the output. This generates a double sideband suppressed carrier signal. In other words the carrier, being one of the input signals, is suppressed by this circuit.

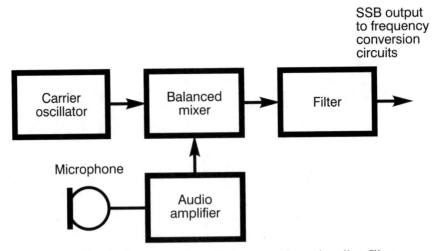

Figure 6.9 *Block diagram of an SSB generator using the filter method*

Figure 6.10 *An amateur SSB transceiver (Courtesy of Yaesu UK Ltd)*

Once this signal has been generated, it is passed into a filter to remove the unwanted sideband. The filter will typically have a selectivity sufficient to pass the required audio bandwidth. For most purposes this will be communications quality only and the bandwidth will be about 2.7 kHz or slightly more.

The position of the carrier frequency relative to the filter pass-band is important. It is necessary to ensure that the wanted sideband is not attenuated while ensuring that no undue amounts of the unwanted sideband are passed through. Normally the filter and carrier are positioned as shown in Figure 6.11.

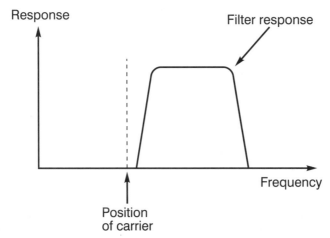

Figure 6.11 *Frequency of carrier and filter relative to one another*

It is normally found that the filter will give additional attenuation to the carrier, enabling high degrees of attenuation than the mixer alone can provide. For most transmitters the carrier will be suppressed by at least 40 dB as will the unwanted sideband.

The single sideband signal is generated on a fixed frequency. Most transmitters will need to be used on a variety of frequencies. To enable this to be achieved the superhet principle is used in the reverse way to that employed in a receiver. The fixed frequency single sideband signal is mixed with a variable frequency local oscillator to give a mix product on the required frequency. As in the case of a receiver there are a number of methods of achieving this. A single variable frequency oscillator on its own is unlikely to give sufficient stability. Some older transmitters may use a crystal mixer oscillator system, or two conversions may be used, the first controlled by a relatively low frequency oscillator, and the second using a switched higher frequency crystal oscillator. Alternatively those transmitters requiring a limited number of channels had the option of using a single crystal oscillator with different crystals switched into circuit to give the required channel. However, today, virtually all sets use a frequency synthesizer to give the stability and coverage without the need for band switching.

When mixers are used it is very important to remove the unwanted mix products. If not they will pass through the later stages of the transmitter and it will be difficult to remove them all in the final stages of filtering. As such it is good practice to place good filters after each mixer to ensure that the unwanted signals are removed as soon after they are generated as possible.

Apart from converting the signal to its final frequency the signal also needs to be amplified to the required level. In view of the fact that the signal carries analogue modulation the amplifier must preserve the nature of the modulation and must be linear. Any distortion will result in the audio becoming distorted as well as the signal occupying a wider bandwidth and causing interference to others.

The second method of generating a single sideband signal is to use phasing techniques. Although not widely used today phasing held a number of advantages as a cheap and easy method of producing an SSB signal in some applications in the past. A block diagram for this method of SSB generation is shown in Figure 6.12. The audio and carrier are split into components shifted by 90° and applied to balanced modulators. The outputs from these are combined, at which point one sideband is reinforced while the other is cancelled out. The reverse sideband can be selected by transposing the audio or carrier phasing.

By using this method of SSB generation it is possible to generate an SSB signal at the operating frequency. However, it is necessary to maintain very accurate control of the amplitude and phase of the signals if cancellation of the reverse sideband is to be achieved. Even a 1° change in the phase of either the RF or audio signals will reduce the sideband suppression to 40 dB if a

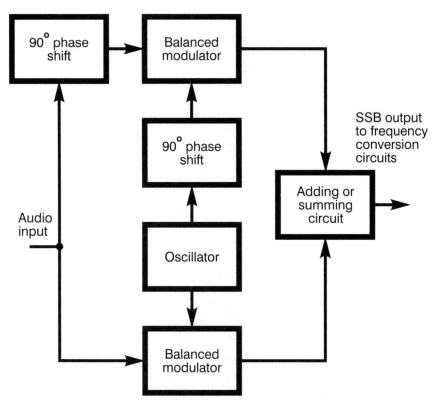

Figure 6.12 *Phasing method of SSB generation*

perfect match of the amplitude of both signals is maintained. Similarly mismatches in the levels will cause the suppression of the unwanted signals to be reduced. This limits the bandwidth of operation quite considerably. The generation of a 90° phase shift over the complete audio bandwidth requires some innovative circuit design. Often high tolerance components are required to achieve this. Another solution is to use a Gingell polyphase shift network. Using this it is possible to obtain the correct phase shift using standard tolerance components. In view of the ease with which the filter method can be implemented, and the flexibility of frequency synthesizers to give tuning, the phasing method of SSB generation has few advantages to offer. Accordingly it is not widely used today except in some limited applications.

Frequency modulation transmitter

Frequency modulation is used for a variety of applications from wide-band FM broadcasting to hand-held transceivers using narrow-band FM. There are

Figure 6.13 *A hand-held FM transceiver*

a variety of circuits for producing an FM signal, and these depend on the type of modulation, whether it is narrow- or wide-band and the circuits being used.

Often FM transmitters take the form shown in Figure 6.14. Here the modulating signal is applied directly to the oscillator. This can be accomplished as shown in Figure 6.15. The audio signal applied to the circuit changes the capacitance of the varactor diode. As the crystal is operating in its parallel resonant mode its frequency can be changed by external components. In this way the frequency of oscillation can be changed, if even by a relatively small amount.

FM transmitters are usually used for frequencies above 30 MHz and as a result crystals are normally needed to provide sufficient stability. The usual way in which these transmitters work is for the oscillator to run at a low

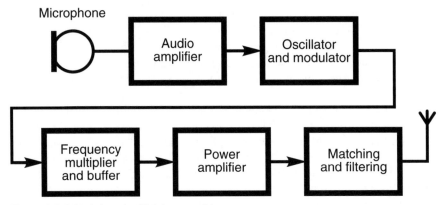

Figure 6.14 *A basic FM transmitter*

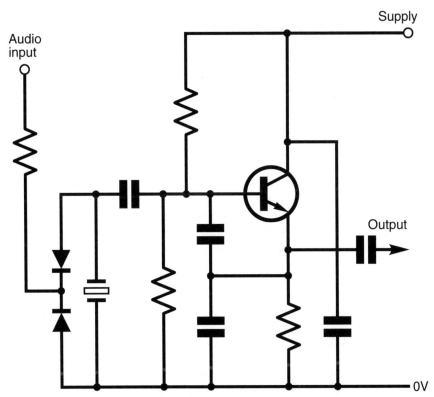

Figure 6.15 *Applying frequency modulation to an oscillator*

frequency, typically a few megahertz, and this signal is multiplied in frequency by a series of multipliers. These are normally amplifiers running in class C to produce high levels of harmonics. Tuned circuits in the output circuit select the correct harmonic. Normally these multipliers are restricted to relatively low multiplication factors, two and three being the most common.

It is found that as the frequency is multiplied so the level of deviation is also increased. This means that if a signal is multiplied 18 times before being transmitted, the level of deviation applied to the oscillator only needs to be relatively small. For a final deviation of 3 kHz, the deviation needs to be 3/18 = 167 Hz for a multiplication factor of 18. This can be achieved relatively easily using the circuit shown.

Applying audio directly to the oscillator is not the optimum method of generating FM. The level of FM can vary from one crystal to the next if the specifications of all the crystals are not exactly the same. Also the linearity of the conversion is not always good. This can be improved by applying a fixed bias to the varactor diode to overcome its non-linearities. However, a superior method is to use a phase modulator placed after the oscillator, as shown in Figure 6.16. The actual frequency deviation given by the phase modulator increases with the increasing audio frequency at a rate of 6 dB per octave. The phase modulator can be made to give an FM compatible signal altering the audio response of the audio amplifiers to give the inverse response. The use of a phase modulator of this nature is far more satisfactory than modulating the oscillator. Sometimes this method of modulation is known as indirect FM as shifting the phase of a signal gives a corresponding change in frequency.

Once the FM has been generated it is amplified (and frequency multiplied) to the correct level (and frequency). Class C amplifiers can be used because all the information for the modulation is contained within the frequency changes. There should not be any amplitude variations. The filtering stages are particularly important for this type of transmitter where a large number of harmonics are generated. Each stage should contain sufficient filtering to ensure that the levels of spurious signals are kept to an acceptable level. Then at the output the filtering and matching circuit must ensure a good match to the antenna system as well as keeping any unwanted signals from being radiated.

Speech processing

When using SSB and AM the level of audio modulation governs the readability of the signal at the receiver. An AM carrier with low modulation will have a lower readability than one of the same power with full modulation. Similarly an SSB signal is dependent upon the level of audio for the level of power transmitted.

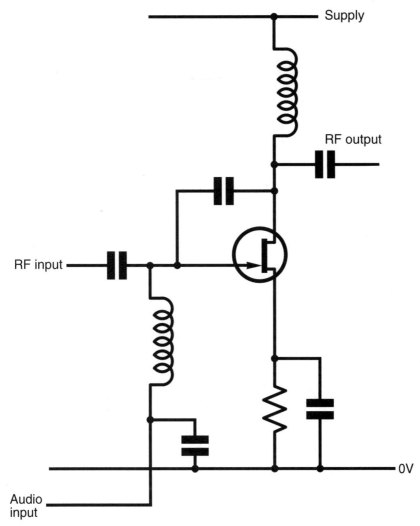

Figure 6.16 *A phase modulator circuit*

Most audio waveforms have a very low duty cycle. In other words their average level is very low when compared to the peak level, as shown in Figure 6.17. As the transmitter must be able to cope with the peak level without overloading the average level of modulation would be low. In turn this results in the signal appearing to be weaker than might be expected. To counteract this most transmitters incorporate some form of speech processing. This has the effect of increasing the average level of modulation, thereby increasing the readability of the signal.

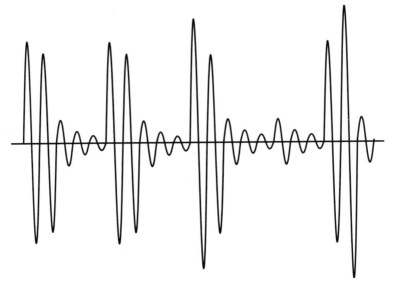

Figure 6.17 *A typical audio waveform*

There are a number of ways in which this can be achieved. The most straightforward is to compress the dynamic range of the signal. If a person is speaking it is found that the average level of the audio varies as more emphasis is placed on some words than others. Also some syllables in the words have a much greater level than others. These level variations can be reduced by using an amplifier which has a controllable gain. The gain can be reduced during periods when the audio level is high and increased when the audio level is low. This type of circuit is known as a compressor as it compresses the dynamic range of the signal. Sometimes these circuits are also called Voice Operated Gain Adjusting Devices (VOGAD). Normally a time constant is involved so that it adjusts according to the general level of the signal and not the instantaneous level. For it to operate correctly it must have a fast attack time to ensure that transients do not pass through unattenuated and cause overloading and distortion in later stages. The decay time must be slower so that the average level of the signal is accommodated as shown in Figure 6.18, and the gain is not continually being adjusted by large amounts.

The compressor does nothing to increase the average level within a particular sound. Even a constant sound where the speaker is not emphasizing words or the level of the sound is not varying for other reasons still has a very low average power level. This can be increased by removing the peaks as shown in Figure 6.19.

Clipping a signal in this way introduces distortion in the form of intermodulation and harmonic distortion. Naturally many of these products

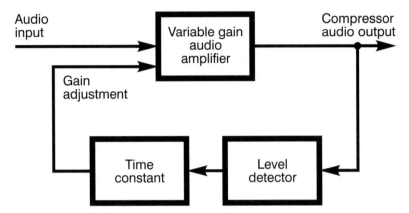

Figure 6.18 *Action of a compressor*

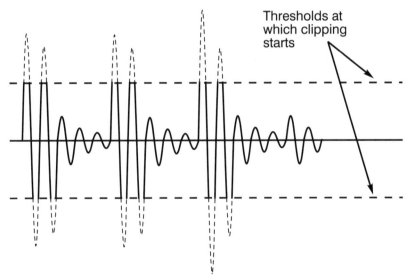

Figure 6.19 *Clipping an audio signal*

fall outside the normal bandwidth, harmonics being at twice, three times, etc. the frequency. These can be removed by using a low pass filter with a cut-off frequency equal to the maximum audio frequency required. This is applied immediately after the clipper.

Despite the filter, distortion remains inside the required audio bandwidth as shown in Figure 6.20. However, for communications applications it provides a significant increase in the average power level which can more than make up for the degradation in intelligibility provided that the level of

Amplitude

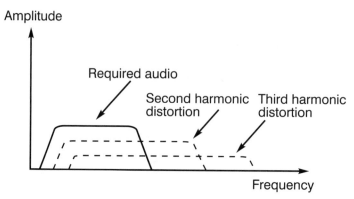

Figure 6.20 *Distortion products from AF clipping*

clipping does not become too much. The level of clipping is defined as the ratio of the peak level without clipping to the peak level with clipping. Normally a level of clipping of around 12 to 15 dB is about the maximum, and this often gives a gain of around 3 or 4 decibels.

To overcome the problem of in-band distortion it is possible to use RF clipping. Using this technique a single sideband signal is generated and then clipped. Harmonics of the signal fall at twice and three times the radio frequency and can be removed very easily. Once clipped and filtered the audio can be regenerated. As the levels of distortion are much lower the amount of clipping which can be applied is much greater. In theory infinite clipping can be applied, but in practice levels of up to 30 dB are used. This can give gains of the order of 10 to 15 dB. With the high levels of clipping, the intelligibility is maintained, but the tone of the sound is changed.

Amplitude

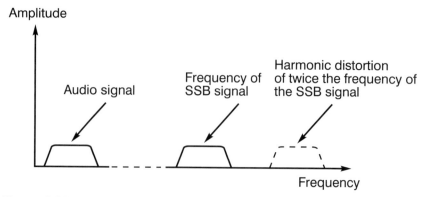

Figure 6.21 *Distortion products from RF clipping*

Broadcast stations also use audio processing. A number of systems are in use dependent upon the type of modulation and the programme content. These systems are often very sophisticated to enable the maximum amount of modulation to be carried without reducing the fidelity of the transmission to an unacceptable level. FM stations rely heavily on compression. Stations carrying popular music apply far more compression so that they keep a 'bright' sound which will attract and keep listeners. Particular attention has to be paid to the fact that many listeners will be in automobiles and they will not want to be continually adjusting the volume control on the set as the levels vary. Stations carrying classical music are far more aware of trying to preserve the dynamic range of the original tracks, and know that most of their listeners are not listening on high fidelity systems but many are in automobiles and will need a reduced dynamic range if they are to hear the music. Some stations even change the level of processing at different times of the day to cater for the different types of listener.

Stations using AM often employ different techniques. Those on the short wave bands are particularly aware of the needs of processing. Some use forms of upward modulation, as well as varying the carrier level to accommodate the changes in level of modulation. This can produce significant savings in power consumption as the carrier level can be reduced in times of low modulation, thereby making major savings in cost. A significant factor when the transmitter may be running at nearly a megawatt.

7 Antenna systems

The antenna system is an important part of any radio system. Its performance will often determine how well the system as a whole can perform. The choice of the correct antenna, as well as its siting and installation, can often give improvements equivalent to raising the transmitter power many times. As a result it is necessary to invest wisely in the antenna system.

The function of the antenna is to pick up the radio electromagnetic waves and convert them into electrical signals. Once they exist as electrical signals they are transferred from the antenna element itself into the receiver where they are amplified, filtered and demodulated to give the required audio output. Conversely in terms of a transmitter the antenna performs the function of transferring the electrical energy into radiated electromagnetic signals.

Antennas have a variety of properties. They can only operate efficiently over a given bandwidth. They also have an electrical impedance, and they are polarized, only picking up waves of a certain polarization. Many of these factors are very important and if the antenna is to give the optimum performance it is necessary to ensure it has the correct properties and that it is set up correctly.

Resonance and bandwidth

An antenna is a form of tuned circuit which acts like an inductor at some frequencies and a capacitor at others. Like a tuned circuit it has a resonant frequency, and most antennas are operated at resonance. In view of this there is only a limited band over which the antenna operates efficiently.

The actual frequency at which the antenna resonates is governed chiefly by its length. The length of the antenna or its elements are generally approximate multiples of a quarter wavelength as the designs of specific antennas will

Figure 7.1 *The antenna system used by the BBC at their Asia relay station in Thailand (Courtesy of the BBC)*

show. This means that the lower the frequency of operation the larger the antenna is likely to be.

The bandwidth is particularly important where transmitters are concerned. If the transmitter is operated outside the bandwidth of the antenna, it is possible that damage may occur. In addition to this the signal radiated by the antenna may be less for a number of reasons.

For receiving purposes the performance of the antenna is less critical in some respects. It can be operated outside its normal bandwidth without any fear of damage to the set. Even a random length of wire will pick up signals, and it may be possible to receive several distant stations. However, for the best reception it is necessary to ensure that the performance of the antenna is optimum. Often good antenna systems will enable a receiver to pick up stations at good strength which are totally inaudible on a receiver with a poor antenna.

Impedance

In the same way that an electrical circuit has a certain impedance, the same is true for an antenna. Each antenna has a certain inductance and capacitance, and for any given frequency the antenna will have a certain impedance. At resonance the inductance and capacitance cancel one another out, leaving a purely resistive element. This is known as the radiation resistance, and varies from one type of antenna to another.

When a resistor is present in a traditional circuit, power is dissipated when a current flows. The same is true for radiation resistance of the antenna, as it represents the power lost due to the radiation of the signal. In fact if a signal generator is connected to the antenna it would see the same resistive load as if a resistor of the same value was present. In view of this transmitters are often run into resistive loads when they are being tested, and no power is to be radiated. In this way the transmitter sees a load equivalent to the antenna.

Most antennas are designed to give a specific resistance. Domestic systems for television and hi-fi are 75 ohms whereas commercial, CB, and amateur radio systems are normally 50 ohms.

The value of the antenna impedance is very important if the system is to operate efficiently. In order to obtain the maximum power transfer into or out of the antenna the transmitter or receiver should have the same value of resistance. This value of resistance can be calculated using Ohm's law and by knowing the values of current and voltage at the feed point of the antenna.

Gain and directivity

One important characteristic of an antenna is the way in which it is more sensitive to signals in one direction than another. This is called the directivity of the antenna. To explain some of the features about directivity it is easier to visualize the operation of the antenna when it is transmitting. The antenna will then be found to perform in the same manner when it is receiving.

Power delivered to the antenna is radiated in a variety of directions. The antenna design may be altered so that it radiates more in one direction than another. As the same amount of power is radiated, it means that more power is radiated in one direction than before. In other words it appears to have a certain amount of gain over the original design.

In order to see the directional pattern of an antenna a polar diagram is plotted. This is a plot of the signal strength around the antenna with the distance of the line away from the antenna indicating the relative strength. A simple dipole antenna may have a pattern like that shown in Figure 7.2. From this it can be seen that the maximum radiation for transmission, and hence the maximum sensitivity to received signals, occurs when the signal is at right angles to the antenna.

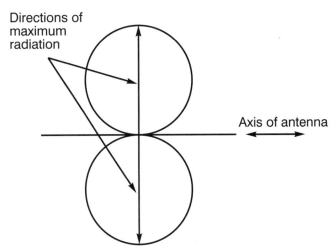

Figure 7.2 *Polar diagram of a half wave dipole*

When an antenna is designed to be directive, its polar diagram may look more like that shown in Figure 7.3. In this diagram it can be seen that the antenna radiates far more signal in one direction than another. The fact that it 'beams' the power in a particular direction means that antennas of this nature are often called beams.

The region of maximum radiation is called the major lobe. However, there are other areas around the antenna where there are significant levels of radiation. These are called minor lobes. They are always present to some extent, and generally the largest is in the opposite direction to the main lobe.

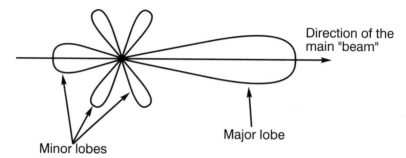

Figure 7.3 *Polar diagram of a directive antenna or beam*

Sometimes the front to back ratio is of interest, and a front to back ratio is quoted and expressed in decibels. This is simply the ratio of the signals in both directions expressed in decibels. It is of particular interest in situations when there are likely to be significant levels of interference in the opposite direction to a wanted signal.

One of the major design parameters of any beam antenna is its gain. This has to be compared to another antenna. The most common antenna used for comparisons is called a dipole. The gain is simply the ratio of the signal from the beam antenna compared to the dipole expressed in decibels.

Sometimes another type of antenna may be used. Called an isotropic source it is an imaginary antenna that radiates equally in all directions. It can be calculated that a dipole has a gain of 2.1 dB over an isotropic source, and therefore when any antennas are quoted against an isotropic source the gain is measured against a dipole and then 2.1 dB are added.

It is necessary to specify what the gain of the antenna is being compared against. The gain of an antenna over a dipole is quoted as a certain number of dBd (dB gain over a dipole). Similarly figures against an isotropic source are quoted as dBi (dB gain over an isotropic source).

Angle of radiation

An antenna like a horizontal dipole will radiate power at all angles relative to the earth, ignoring the effects of the ground nearby. This means that some power will travel vertically upwards whereas some power will travel almost horizontally along the ground. Whether the antenna is horizontal, or vertical like that shown in the diagram, the angle of radiation is very important.

Distant signals on short wave frequencies usually arrive at the antenna at a low angle. This means that if distant stations are being sought it is necessary to use an antenna with a low angle of radiation. Having an antenna like this also helps reduce interference from less distant stations as they are usually received at a much higher angle. Vertical antennas generally have a low angle

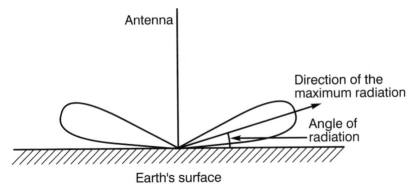

Figure 7.4 *Angle of radiation*

of radiation, and are much less sensitive at higher angles, making them ideal for long distance work. For VHF and UHF operation signals must have a low angle of radiation as signal paths follow the earth's surface.

Height

The height of an antenna is very important. If it is installed close to the ground then it will be shielded by nearby objects, reducing the overall signal levels, especially the low angle long distance signals. Raising the height of the antenna will not only enable the antenna to see over these objects, but it also increases the distance to the horizon, again increasing the coverage of the antenna. Raising the height of the antenna will improve its operation at all frequencies, whether on the short wave frequencies or at VHF and above.

Antenna system

It is not possible to consider the antenna elements in isolation. It is necessary to look at the complete antenna system and optimize the operation of the whole system. The antenna system can be split into three main parts. The antenna elements themselves, the feeder, and the matching unit providing the correct match between the antenna and the feeder. This is used to transfer the power from the place where the antenna itself is located into the receiver.

Feeder

The purpose of a feeder is to carry radio frequency signals from one point to another with the minimum amount of signal loss. In view of the fact that radio frequency signals are being carried, ordinary wire like that used for carrying mains power is unsuitable. Feeders with suitable radio frequency

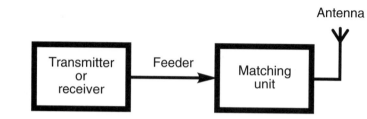

Figure 7.5 *An antenna system*

characteristics are required. A poor feeder may result in the whole antenna system being degraded. It is therefore very important that a suitable feeder is used.

The operation of a feeder may not be as straightforward as might be expected at a first look. There are a number of characteristics which need to be taken into consideration.

Feeder impedance

One of the first aspects of a feeder to note is its impedance. Just as an antenna has a certain value of impedance, and a receiver or transmitter has an input or output impedance, a feeder has what is called its characteristic impedance. This is particularly important because this has to be matched to the value of impedance for the transmitter or receiver and the antenna if the optimum performance is to be achieved.

The impedance of the feeder is governed by a number of factors. The actual dimensions have a large bearing on the impedance, as does the dielectric constant of the medium in and around the feeder. By controlling these factors it is possible to manufacture feeders of the required impedance.

Standing waves

When the whole antenna system is perfectly matched the maximum power transfer is obtained. When this does not happen, which is the case in real systems to a certain degree, standing waves are set up in the feeder.

When power is transferred from a source into the load, the maximum power transfer occurs when the load and source have the same impedance and are said to be matched. In the case of a feeder and an antenna, the feeder acts as the source and the antenna is the load. If there is a poor match only a proportion of the power will be transferred from the feeder into the antenna. The remaining power from the feeder cannot just disappear and is reflected back along the feeder. When this happens the voltages and currents in the

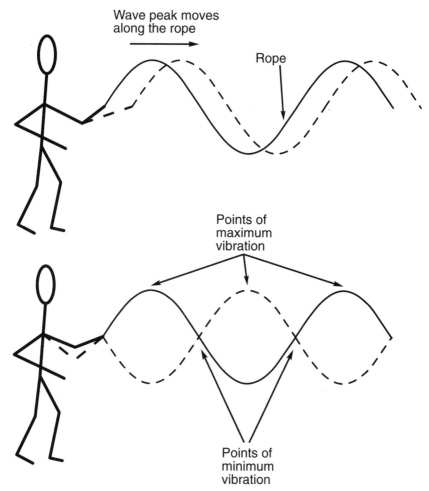

Figure 7.6 *Analogy of standing waves*

feeder add and subtract at different points along the feeder. The result of this is that standing waves are set up.

The way in which the effect occurs can be demonstrated with a length of rope. If one end is left free and the other is moved up and down as shown in Figure 7.6, the wave motion can be seen to move along the rope. However, if one end is fixed a standing wave motion is set up, and points of minimum and maximum vibration can be seen.

When the feeder and load are perfectly matched the current and voltage will be constant along the feeder, as shown in Figure 7.7.

If the load impedance does not match that of the feeder a discontinuity is created. The feeder wants to supply a certain voltage and current ratio, while

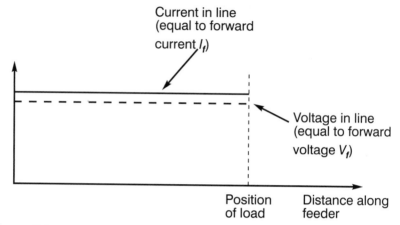

Figure 7.7 *Voltage and current magnitude along a perfectly matched line.*

the load must also obey Ohm's law as well and cannot accept the same voltage and current ratio. To take an example a 50 ohm feeder with 100 watts entering will have a voltage of 70.7 volts and a current of 1.414 amps. A 25 ohm load would require a voltage of 50 volts and a current of 2 amps to dissipate the same current. To resolve this discontinuity, power is reflected and standing waves are generated.

When the load resistance is lower than the feeder impedance, voltage and current magnitudes like that shown in Figure 7.8 are set up. Here the total current at the load point is higher than that of the perfectly matched line, whereas the voltage is less.

The values of current and voltage along the feeder vary as shown along the feeder. For small values of reflected power the waveform is almost sinusoidal, but for larger values it becomes more like a full wave rectified sine wave. This waveform consists of voltage and current from the forward power plus voltage and current from the reflected power. At a distance a quarter of a wavelength from the load the combined voltages reach a maximum value while the current is at a minimum. At a distance half a wavelength from the load the voltage and current are the same as at the load.

A similar situation occurs when the load resistance is greater than the feeder impedance; however, this time the total voltage at the load is higher than the value of the perfectly matched line. The voltage reaches a minimum at a distance a quarter of a wavelength from the load and the current is at a maximum. However, at a distance of a half wavelength from the load the voltage and current are the same as at the load.

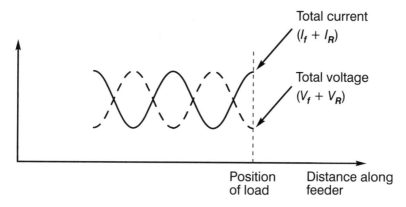

Total current
$(I_f + I_R)$

Total voltage
$(V_f + V_R)$

Position
of load

Distance along
feeder

(a) Load resistance is lower than feeder impedance

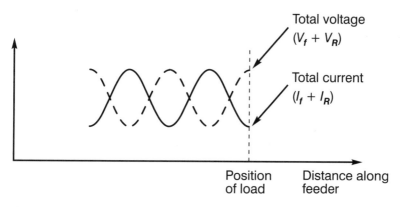

Total voltage
$(V_f + V_R)$

Total current
$(I_f + I_R)$

Position
of load

Distance along
feeder

(b) Load resistance is higher than feeder impedance

Figure 7.8 *Voltage and current magnitudes for a mismatched line*

It is often necessary to have a measure of the amount of power which is being reflected. This is particularly important where transmitters are used because the high current or voltage values may damage the feeder if they reach very high levels, or the transmitter itself may be damaged. The figure normally used for measuring the standing waves is called the standing wave ratio (SWR), and it is a measure of the maximum to minimum values on the line. In most instances the voltage standing wave ratio (VSWR) is used.

The standing wave ratio is a ratio of the maximum to minimum values of standing waves in a feeder. The reflection coefficient (ρ) can be defined and this is the ratio of the reflected current or voltage vector to the forward voltage or current vector. It is therefore very easy to calculate the SWR. The minimum value of standing wave is $(1 - \rho)$ and the maximum is $(1 + \rho)$. The standing wave ratio then becomes:

$$SWR = \frac{(1 + \rho)}{(1 - \rho)}$$

From this it can be seen that a perfectly matched line will give a ratio of 1:1 while a completely mismatched line gives ∞:1. Although it is perfectly possible to quote VSWR values of less than unity, it is normal convention to express them as ratios greater than one.

Even though the voltage and current vary along the length of the feeder, the amount of power remains the same if losses are ignored. This means that the standing wave ratio remains the same along the whole length of the feeder.

Often the forward and reflected power may be measured. From this it is easy to calculate the reflection coefficient as given below:

$$\rho = \sqrt{\frac{P_{ref}}{P_{fwd}}}$$

where P_{ref} is the reflected power
P_{fwd} is the forward power

Loss

Another important factor about a feeder is the loss which it introduces into the system. The ideal scenario would be for the same amount of power to appear at the far end as entered from the generator. In reality the radio frequency power leaving the far end of the cable is always less than that which entered.

There are a number of reasons for this. The first is that the conductors in the feeder have a certain resistance, and as a result some of the power is dissipated as heat. To reduce this the conductors can be made thicker, but this increases the size of the whole cable, and increases the cost.

Another reason for power loss arises from the dielectric. A dielectric material is used between conductors in the feeder to act as an insulator and spacer to keep conductors a given space apart. Poor quality dielectric can dissipate some power. It is for this reason that many dielectrics are semi-air spaced, consisting of plastic with air holes of one variety or another. Moisture entering the feeder dielectric can increase the loss dramatically. As a result it is necessary to ensure that any feeders which are used externally need to be well sealed where required.

Finally some power can be lost by radiation. In many types of feeder the amounts of power lost by radiation are relatively low. However, the reverse effect can be serious in some instances. If the feeder runs through an area where noise levels are high it may pick up significant amounts of interference which could adversely affect reception, despite having an antenna in a location where interference local levels are low.

The loss introduced by a certain type of feeder is proportional to its length. As a result figures for this type of specification are given for a given length of feeder, often 10 metres. They are also expressed in decibels. The frequency of operation also has a bearing, the loss rising with frequency. As a result a typical specification for a feeder might be that it has a loss of 1.0 dB at 100 MHz and 3.8 dB at 1000 MHz for a 10 metre length. For the best operation of the antenna it is necessary to ensure that the loss is minimized in keeping with the cost of installing the feeder.

Velocity factor

When a signal travels in free space it travels at the speed of light. It would also travel at the same speed in a feeder if it did not contain an insulating dielectric. This is normally used to space the wires in the feeder and may be a form of plastic, or possibly even just air. The speed of the signal is reduced by a factor of $1/\sqrt{\varepsilon}$ where ε is the dielectric constant. The dielectric constant is always greater than one, and as a result the speed of the signal is always less than the speed of light. For many coaxial types of feeder the velocity factor is around 0.66 (i.e. 0.66 times the speed of light) although in some cases it can be as low as 0.5 and higher in others. For open wire type feeders the velocity factor may be much greater at around 0.98.

Apart from reducing the velocity of the signal, the wavelength also changes and is reduced by the same factor. This results from the fact that the lower speed means that it travels a shorter distance in the same time. In many cases this does not cause a problem, but for applications where the feeder is cut to a specific number of wavelengths, this can be crucial.

Types of feeder

There are a number of different types of feeder which can be used dependent upon the requirements of the application. Some types are most suitable for more specialist or exacting requirements whereas other types like the very familiar coaxial feeder provide good performance in a wide variety of areas.

Coaxial feeder

This type of feeder is the most commonly used, being found in most homes for television or VHF FM down leads. Apart from this it is widely used for very many other applications where a radio frequency feeder is required.

Coaxial feeder consists of two concentric conductors, spaced apart from one another by an insulating dielectric as shown in Figure 7.9. The inner conductor may be single or multi-stranded, but the outer one normally consists of a braid as shown in the diagram. The outside of the cable is covered by a protective sheath to prevent moisture ingress, as well as providing some mechanical protection.

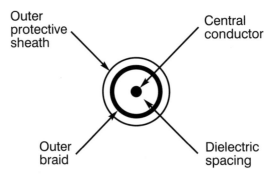

Figure 7.9 *Cross-section through a piece of coaxial cable*

Coax is the most commonly used type of feeder. It has a number of advantages, including the fact that it is not affected by nearby objects. This means that it can be run almost anywhere without any ill effects. Although it will pick up or radiate very small amounts of signal this is normally small enough to be ignored. Where superior screening is required, the screen can be made up from a double layer of braid. In some instances it may be made from solid copper, although this makes it very difficult to bend.

The cable carries currents in both the inner and outer conductors. As they are equal and opposite they cancel one another out and all the fields are confined to the cable. It is for this reason that its operation is not affected by the proximity of nearby objects.

There are a number of different types of coax. The main distinguishing feature is the impedance. For most television and domestic hi-fi antennas 75 ohms has been adopted as standard. For commercial applications, amateur radio and citizens' band equipment 50 ohms is used as the standard.

The impedance is determined by the dimensions of the conductors and the dielectric constant of the material between them. It can be calculated from the formula:

$$Z_o = \frac{138}{\sqrt{\varepsilon}} \quad \log_{10}\left(\frac{D}{d}\right)$$

where D is the inside diameter of the outer conductor or braid

d is the outer diameter of the inner conductor

ε is the dielectric constant of the material between the two conductors

Apart from the loss of a feeder which has already been covered, the other aspect of a coaxial feeder to note is that it is what is called an unbalanced feeder. This means that one of the conductors is connected to earth. As one would expect the outer braid is always connected to earth as it acts as a screen. The fact that coax is unbalanced means that it must be used with a system which can tolerate unbalanced feeders. This is normally no problem with transmitters and receivers, but some antennas require a balanced feed. In cases like this a balun must be used. This is essentially a radio frequency transformer used to isolate the signals from a direct connection to earth.

Balanced feeder

In some instances balanced feeders need to be used in antenna systems. Often called twin or open wire feeder this type of feeder is not nearly as widely used as coax but has a number of advantages in some applications.

A balanced or twin feeder consists of two parallel conductors as shown. The currents flowing in both wires run in opposite directions but are equal in magnitude. As a result the fields from them cancel out and no power is radiated or picked up. To ensure efficient operation the spacing of the conductors is normally kept to within about 0.01 wavelengths.

This type of feeder can take a variety of forms. An 'open wire' feeder can be made by having two wires running parallel to one another. Spacers are used every 15 to 30 centimetres to maintain the wire spacing. Usually these are made from plastic or other insulating material. The feeder may also be bought as a 300 ohm ribbon consisting of two wires spaced with a clear plastic. This is often used for temporary internal VHF FM antennas. It can also be bought with a black plastic dielectric with oval holes spaced at intervals. This type gives a better performance than the clear plastic varieties which absorb water if used outside.

Like coaxial cable, the impedance of twin feeder is governed by the dimensions of the conductors, their spacing and the dielectric constant of the

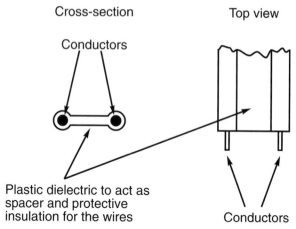

Figure 7.10 *Twin feeder*

material between them. The impedance can be calculated from the formula given below.

$$Z_o = \frac{276}{\sqrt{\varepsilon}} \quad \log_{10}\left(\frac{D}{d}\right)$$

where D is the distance between the two conductors
   d is the outer diameter of the conductors
   ε is the dielectric constant of the material between the two conductors

Open wire or twin feeder is not nearly as widely used as coax, although it provides an ideal solution for a number of applications, especially those in the short wave part of the spectrum. It has a velocity factor of about 0.98 when the open wire version is used, and can offer very low levels of attenuation if it is kept away from other objects. The main drawback is that it is affected by nearby objects and as a result it cannot be taken through buildings in the same way as coax. This limits its use considerably, and it is rarely seen in domestic applications except for use with temporary VHF FM antennas.

Waveguide

A third major type of feeder is called a waveguide. This is only used for microwave frequencies and it consists of a hollow pipe along which the signals propagate. Waveguides can be circular, but it is more common to see rectangular types as shown in Figure 7.11. They are different to the more

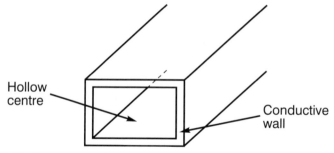

Figure 7.11 *A rectangular waveguide*

conventional forms of feeder in that there are not two conductors. Instead the signal is introduced into the waveguide and this carries it as far as required, the walls preventing the signal escaping.

A signal can be entered into the waveguide in a number of ways. The most straightforward is known as a launcher. This is basically a small probe which penetrates a small distance into the centre of the waveguide itself, as shown. Often this probe may be the centre conductor of the coaxial cable connected to the waveguide. The probe is orientated so that it is parallel to the lines of the electric field which are to be set up in the waveguide. An alternative method is to have a loop which is connected to the wall of the waveguide. This encompasses the magnetic field lines and sets up the electromagnetic wave in this way. However, for most applications it is more convenient to use the open circuit probe. These launchers can be used for transmitting signals into the waveguide as well as receiving them from the waveguide.

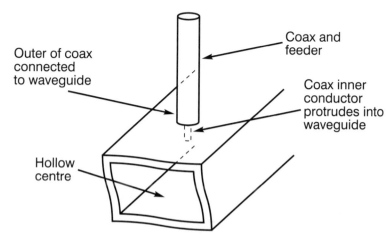

Figure 7.12 *A waveguide launcher*

The dimensions of a waveguide are very important. It is found that below a given frequency called the critical frequency, a waveguide of certain dimensions will not operate. Also if the waveguide is made too large its cost will be higher than necessary and there is the possibility of higher order modes travelling along it which may introduce problems. As a result there are a variety of standard sizes of waveguide, and the correct one can be chosen dependent upon the frequency in use. They are allocated numbers and as an example waveguide WG10 is used for frequencies between 2.60 and 3.95 GHz. The advantage of a waveguide is that it offers a very low degree of attenuation or loss at these frequencies. At these frequencies the loss for a 30 metre length would only be about 1 dB dependent upon the exact frequency and whether the waveguide walls were made from aluminium or copper. This particular waveguide has internal dimensions of 72 × 34 mm and has a cut-off frequency of 2.08 GHz.

Signals can be transmitted directly out of a waveguide into free space. Although the directional properties are not very good there is no need to terminate them in an antenna for them to be able to radiate a signal. As a result it is very important NEVER to look down a waveguide as it is possible that power could be radiated from it if it is connected to a source of radio frequency power. High levels of power can quickly damage the eye.

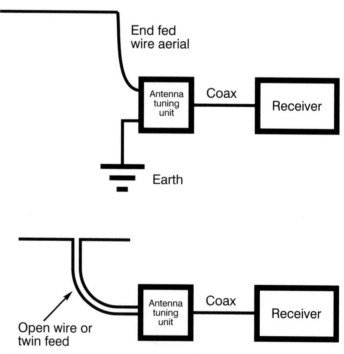

Figure 7.13 *Use of an antenna matching unit*

Matching unit

In order that the maximum amount of signal is transferred from the antenna to the feeder, the impedances must be the same. Unfortunately the impedance of the antenna will change with frequency, and even though it may be designed to give a good match at one frequency, this may not be the case at another. Also the antenna design may be such that it does not match the feeder. This may be the case when the antenna is shorter than a quarter wavelength.

To ensure that the best match is obtained an antenna tuning or matching unit can be used in many instances. There are a number of designs for these units which contain coils and capacitors (fixed and variable) and in some instances resistors may be used so that the antenna can give a good match to the feeder. In this way the antenna can be made to operate at its greatest efficiency and without large levels of VSWR present.

Types of antenna

There is a wide variety of different types of antenna which can be used. Each has its own advantages and may be used for specific applications. Often smaller antennas are not as efficient and are used where portability is of importance, whereas the much larger systems are employed where performance is of paramount importance and cost and size are not the major requirements.

Ferrite rod antennas

The ferrite rod antenna is almost universally used in portable transistor broadcast receivers as well as many hi-fi tuners where reception on the long, medium and possibly the short wave bands is required.

As the name suggests the antenna consists of a rod made of ferrite, an iron-based magnetic material. Coils are wound around this as shown in Figure 7.14. This coil is brought to resonance using a variable tuning capacitor contained within the radio circuitry itself and in this way the antenna can be tuned to resonance. As the antenna is tuned it usually forms the RF tuning circuit for the receiver, enabling both functions to be combined within the same components, thereby reducing the number of components and hence the cost of the set.

The ferrite rod antenna operates using the high permeability of the ferrite material to 'concentrate' the magnetic component of the radio waves as shown in Figure 7.15. This means that the antenna is directive. It operates best only when the magnetic lines of force fall in line with the antenna. This occurs when it is at right angles to the direction of the

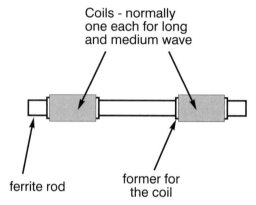

Coils - normally
one each for long
and medium wave

ferrite rod

former for
the coil

Figure 7.14 *A ferrite rod antenna*

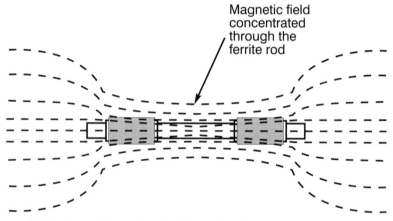

Magnetic field
concentrated
through the
ferrite rod

Figure 7.15 *Operation of a ferrite rod antenna*

transmitter. In other words the antenna has a null position where the signal
level is at a minimum when the antenna is in line with the direction of the
transmitter.

These antennas are very convenient for portable applications, but their
efficiency is much less than that of a larger antenna. The performance of
the ferrite also limits their frequency response. Normally they are only
effective on the long and medium wave bands, but they are sometimes
used for short waves as well where their performance is significantly
degraded, mainly because of the ferrite.

Dipole

The dipole is one of the most common forms of antenna. It is widely used on its own, as well as being the basis of a number of other directive antennas.

The most common form of the dipole is the half wave version although it can be any odd multiple of electrical half wavelengths. The half wave version consists of a half wave length section of wire, with the feed point in the centre of the antenna as shown in Figure 7.16.

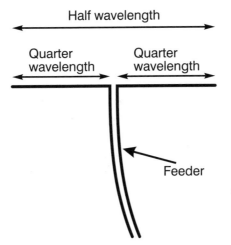

Figure 7.16 *A dipole antenna*

To understand more about the operation of the antenna it is necessary to look at the voltage and current distribution along the wire. From Figure 7.17 it can be seen that they vary sinusoidally along the length of the antenna. The voltage has a maximum at either end, and a minimum in the middle where the feed point is taken. The current follows the opposite pattern, falling to a minimum at the end and having a maximum in the middle. Longer dipoles follow the same basic principles having further wavelengths. However, a voltage maximum and current minimum is found at both ends. The feed point is similarly taken at a point of voltage minimum and current maximum. A three half wavelength antenna is sometimes used, and this could be fed in the middle, or at one of the other current maximum points.

With the feed point taken at the point where the voltage is a minimum and the current a maximum it can be understood from Ohm's law that the impedance is low. The actual impedance depends mainly on the proximity of nearby objects, but in free space the value of the impedance is 78 ohms, making it a good match to 75 ohm coax.

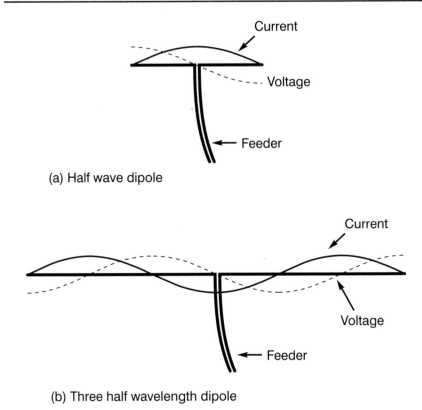

(a) Half wave dipole

(b) Three half wavelength dipole

Figure 7.17 *Voltage and current distribution of dipole antennas*

It is very easy to alter the impedance of a dipole. The proximity of nearby objects has a major effect. The height above ground alters the impedance quite significantly. It is possible to calculate this effect and plot the impedance of a dipole in terms of its height in wavelengths above ground as shown in Figure 7.18.

In some antenna designs the feed impedance of the dipole falls to very low values. Sometimes it may fall to values of 10 ohms or less. This can be very difficult to feed efficiently. The impedance can be increased to more suitable values by using what is called a folded dipole. In this type of antenna an additional wire is taken from one end to the other. Different conductor sizes can be used for the original section of the dipole element and the folded section. If they are both the same then the impedance of the dipole is raised by a factor of four. By changing the ratios of the sizes of the conductors it is possible to tailor the impedance. However, in most cases the same sizes are used, giving a basic impedance of 300 ohms for a folded dipole (i.e. 4 × 75 ohms).

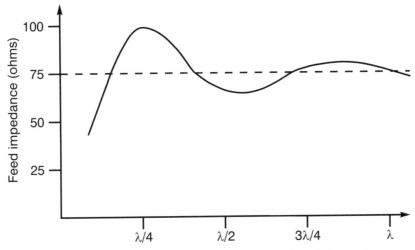

Figure 7.18 *Approximate impedance of a horizontal dipole at varying heights above ground*

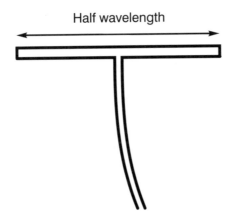

Half wavelength

Figure 7.19 *A folded dipole*

An additional advantage of using a folded dipole is that it gives an increased bandwidth over a basic dipole. This is of great advantage when the antenna needs to be used over a band of frequencies as in the case of VHF FM or UHF television broadcasts.

The length of the dipole is of prime importance because it determines the resonant frequency. This can be determined quite easily although it is not quite the same as a half wavelength in free space. Instead it is slightly shorter.

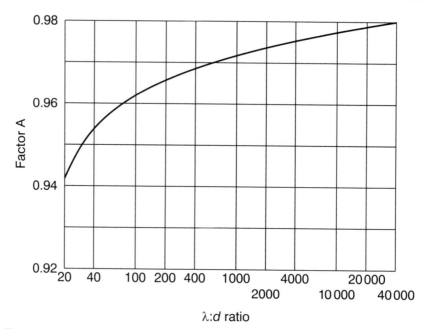

Figure 7.20 *Length reduction of an antenna according to conductor size*

This is due to a number of effects including one referred to as the end effect. This is largely dependent upon the length to diameter ratio of the wire or conductor which is used. In most cases the reduction in length is around 5%, although it is possible to determine a more exact value from the graph in Figure 7.20.

From a knowledge of the length reduction factor it is possible to calculate the antenna length in either metres or inches.

$$\text{length (metres)} = \frac{A \times 150}{f}$$

$$\text{length (inches)} = \frac{A \times 5905}{f}$$

where f is the frequency of operation in MHz

End fed wire

One of the easiest types of antenna to install for short wave applications is an end fed wire. Often misnamed a long wire, this type of antenna can be erected

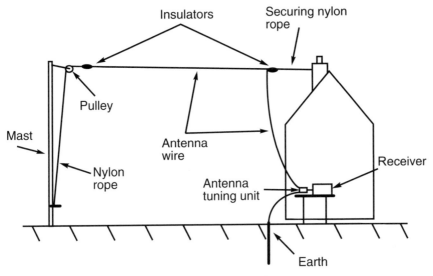

Figure 7.21 *An end fed wire antenna system*

very easily and can be used on a variety of frequencies. As such these antennas are used in a number of applications including ships and aircraft as well as being very convenient for short wave listeners and radio amateurs.

The antenna is normally made to be a quarter wavelength or longer at the lowest frequency of operation, but to give a good match it is necessary to use the antenna with a matching unit. It also needs to be operated with a good earth system. A poor earth will reduce the efficiency of the system.

The directional pattern of the system will depend upon the length of the antenna. A quarter wavelength system will have its maximum sensitivity at right angles to the axis of the wire. Longer ones will have lobes which move progressively towards the axis of the wire itself. If the antenna is made several wavelengths long it becomes what is known as an 'end fire' antenna where the main lobes are almost in line with the axis of the wire. This is what a true long wire is.

Yagi

The yagi is certainly the most popular form of directional antenna. More correctly named the Yagi-Uda after its Japanese inventors it was first outlined in a paper presented in 1928. Nowadays the antenna is widely used where a directional array is required, being almost the only type of antenna used for UHF television reception.

The basic antenna consists of a central boom with the elements mounted to it at right angles as shown in Figure 7.22. The antenna consists of the main

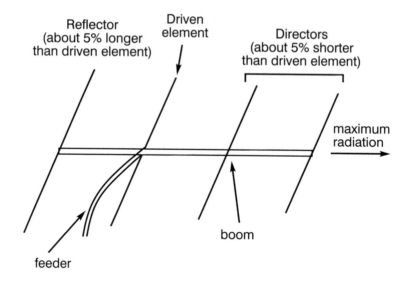

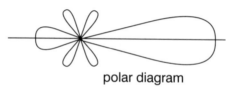

polar diagram

Figure 7.22 *A yagi antenna*

driven element to which the feeder is connected, and parasitic elements either side. These parasitic elements are not directly connected to the feeder but operate by picking up and re-radiating power in such a way that the directional properties of the antenna are altered.

There are two main types of parasitic element: reflectors which reflect power back towards the driven element, and directors which increase the power levels in the direction of the directors. The properties of a parasitic element are determined by their spacing, and their electrical length.

When a parasitic element is made inductive the induced currents are in such a phase that they re-radiate power back towards the driven element, i.e. reflecting the power. To make an element into a reflector in this way it is tuned slightly below resonance. This can be done by physically adding a coil to provide the additional inductance, or more usually by making it physically slightly longer. Typically reflectors are made about 5% longer than the driven elements. Conversely directors are made by tuning the element below resonance. This can be done by adding some capacitance into the element, or making it slightly shorter than the driven element, again by about 5%.

Only one reflector is used, as the addition of further reflectors behind the main one adds very little to the performance. Further directors do give increased directivity, and it is not uncommon to see ten or more directors on an antenna. Normally the number is limited by size, cost and the required amount of gain. Typically a two element design consisting of a driven element and reflector will give around 5 dB gain over a dipole and a three element antenna with a director as well gives around 7 dB. Additional directors give less gain, starting at around 2 dB for the first one or two; however, as a rough guide each additional director gives around 1 dB. For example, a typical 11 element array is likely to have a gain of around 13 dB and a 12 element version, around 14 dB. It is also found that the element spacing has a small effect on the gain. Usually the spacing is between 0.1 and 0.3 wavelengths.

Figure 7.23 *A yagi used for television reception*

The polar diagram of a yagi is generally like that shown in Figure 7.22. As more elements are added the beamwidth becomes much narrower. This makes positioning the antenna more critical, a point to be remembered when choosing the required gain for an antenna.

The feed impedance is altered by the parasitic elements. In the same way that the presence of the earth lowers the feed impedance of the dipole, placing parasitic elements close to the driven element lowers its impedance. The element spacing has a larger effect on the feed impedance than it does on the gain. In most designs a folded dipole is used and the spacing adjusted to give the optimum match.

Vertical antenna

Vertical antennas are widely used, especially at VHF and UHF. As the antenna element is in the vertical plane, power is radiated in all directions around it. Coupled to this the antenna has a low angle of radiation, meaning that little signal is wasted by radiation in an upwards direction away from the earth. In view of their radiation pattern these antennas are ideal for mobile applications. Their all around radiation pattern means that the antenna does not need to be repositioned as the vehicle moves. The use of these antennas is not restricted to mobile stations. Fixed stations too can use them when an omnidirectional pattern is required. Most medium wave broadcast stations use vertical antennas to enable them to cover all directions away from the antenna.

Although vertical antennas can be of different lengths, they are often made a quarter wavelength long. This means that they are fed at a current maximum point as shown in Figure 7.24.

A vertical is an unbalanced antenna and is fed with coaxial feeder as shown. The centre conductor of the feeder is connected to the vertical element, while the screen is connected to a ground. The grounding system must be very good. A single spike into the earth may have a resistance of 100 ohms or more in some cases. As the feed impedance of a vertical of this type

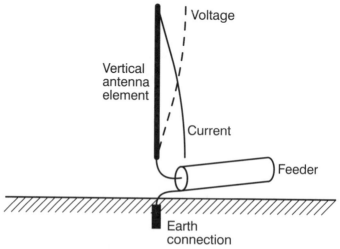

Figure 7.24 *A quarter wave vertical antenna*

may be only 35 ohms, the earth resistance easily becomes the dominating factor absorbing most of the power supplied to the system. Broadcast stations using vertical antennas require the optimum performance and will ensure that a very good earth system is installed along with the antenna itself. Apart from a good earth spike at the base of the antenna radial systems consisting of wires a quarter wavelength or more are buried just below the surface of the ground.

For mobile applications the antenna is mounted onto the bodywork of the vehicle. In this case the metalwork of the vehicle acts as the ground. This is normally a very efficient grounding system, especially at VHF and above where the bodywork extends for several wavelengths away from the base of the antenna. The other advantage is that the antenna can be mounted relatively high up. The best position from a radio performance standpoint is the centre of the roof; however, this is not always acceptable aesthetically. In this case mounting points on the wing usually give good performance as well.

It is not always convenient to mount a vertical on the ground. It can easily be masked by nearby objects reducing its effectiveness. This is particularly true for VHF and UHF applications where the antenna is physically very small and mounting it on the ground would considerably reduce its effectiveness. In situations like this a ground plane system can be used. The ground plane is made out of a number of rods or wires which are normally a quarter wavelength long. Four rods are usually employed as being sufficient for most applications and a ground plane system may look like that shown in Figure 7.25.

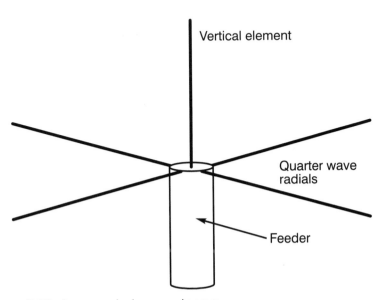

Figure 7.25 *A ground plane antenna*

The feed impedance of a ground plane is low. To improve the match to 50 ohms the radials can be bent downwards. As the radials are bent further down, the antenna becomes more like a vertical dipole and its impedance increases towards a value of 78 ohms, assuming that there are no nearby objects. Alternatively the vertical element can be folded in the same way as a dipole, or a matching network can be used at the feed point to ensure a good match.

Short antennas

It is often necessary to design an antenna which is shorter than a quarter wavelength. With electronic equipment becoming smaller the same is often needed for any associated antennas. This is particularly true for cellular telephones where equipment is very much smaller than it was a few years ago, and also the antennas used on vehicles are required to be smaller.

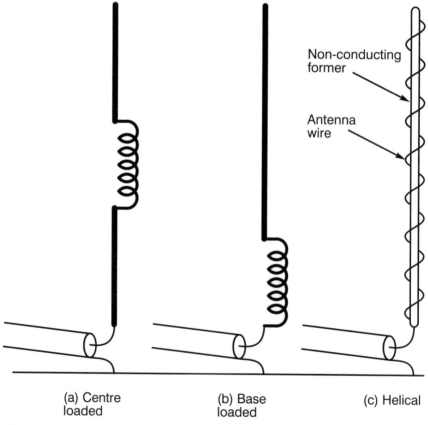

Non-conducting former

Antenna wire

(a) Centre loaded

(b) Base loaded

(c) Helical

Figure 7.26 *Adding inductive loading to an antenna*

There are a number of ways of reducing the physical size of an antenna. One method is to make the antenna electrically the right length even though it is physically small. It is found that when an antenna is too short for its resonant frequency it becomes capacitive. The capacitive component can be removed by adding inductance to the antenna to bring it to resonance. This can be achieved by adding a coil to a section of the antenna as shown in Figure 7.26, or by making what is called a helical antenna and distributing the inductance over the whole length. When a single coil is used to give the inductance, this is either placed in the centre as shown or at the base.

Another alternative is to make the antenna electrically short and feed it accordingly. If this method is adopted arrangements need to be made to match the impedance of the antenna to the feeder, as the two will not be matched.

Parabolic reflector

One type of antenna which has been seen far more in recent years is the parabolic reflector or 'dish'. It is widely used for frequencies above 1 GHz where very high levels of gain can be achieved. Initially these antennas were only used for professional applications, especially radio astronomy or satellite communications. However, with the advent of satellite television these antennas are often seen on the sides of houses for reception of these broadcasts.

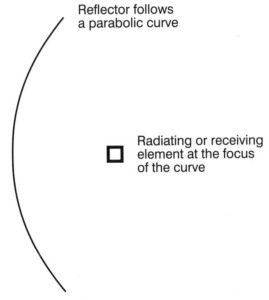

Reflector follows a parabolic curve

Radiating or receiving element at the focus of the curve

Figure 7.27 *A parabolic reflector*

The antenna consists of a radiating element which may be a simple dipole or a waveguide horn antenna. This is placed at the focal point of the parabolic reflecting surface. The energy from the radiating element is arranged so that it illuminates the reflecting surface. Once the energy is reflected it leaves the antenna system in a narrow beam. As a result considerable levels of gain can be achieved.

The gain is dependent on a variety of factors, but is mainly a function of the diameter of the parabolic reflector. The actual gain can be estimated from the formula:

$$G = 10 \log_{10} k \left(\frac{\Pi D}{\lambda} \right)^2$$

where G is the gain over an isotropic source
$\quad\quad k$ is the efficiency factor which is generally about 50%
$\quad\quad D$ is the diameter of the parabolic reflector in metres
$\quad\quad \lambda$ is the wavelength of the signal in metres

From this it can be seen that very large gains can be achieved if sufficiently large reflectors are used. However, when the antenna has a very large gain, the beamwidth is also very small and the antenna requires very careful control over its position. In professional systems electrical servo systems are used to provide very precise positioning.

The reflecting surface may not be as critical as may be thought at first. Often a wire mesh may be used. Provided that the pitch of the mesh is small compared to a wavelength it will be seen as a continuous surface by the radio signals. If a mesh is used then the wind resistance will be reduced, and this may provide significant advantages.

Active antenna

In a number of receiving applications active antennas are used. As the name implies they employ an active or amplifying element. They consist of an electrically short antenna element. An active amplifier at the base of the antenna element is used to amplify the signal and act as a matching circuit to ensure a good match to the feeder. This is normally achieved easily because a short antenna will present a high impedance and this can usually be accommodated very easily with a semiconductor amplifier.

Power for the amplifier is normally supplied along the coax from a power supply at the receiver end of the feeder, the centre conductor of the coax allowing the supply to be passed to the amplifier, while also carrying the signal to the receiver. In this way it is possible to have the power source near the receiver without the need for additional wires to be routed to the antenna.

The design of active antennas is not easy. If they are to perform well they must be capable of good strong signal handling over a wide frequency range. This is necessary because the antenna is untuned and will pick up signals over a wide frequency band, some of which will be very strong. To achieve the required performance devices capable of handling relatively high powers are used. Poor performance will result in intermodulation products being generated, and the receiver seeing many more signals than the receiver element is actually picking up.

8 Broadcasting

Most people first come into contact with radio by listening to radio broadcasts. Today many households have a variety of broadcast receivers from portable sets for use around the home, and car radios to very expensive hi-fi systems. These are used to pick up broadcasts on the long, medium, and short wave bands as well as the higher quality transmissions on the VHF FM band. In addition to this new digital broadcast systems are being introduced and these are capable of providing even higher quality reproduction than that which can be obtained with VHF FM. Apart from providing improved quality these transmissions can also provide a new range of facilities as they are able to transmit data alongside the audio.

AM broadcasts

The first broadcasts to be made used amplitude modulation, and although better methods of modulation are available today, AM is still in widespread use on the long, medium, and short wave bands. There are a number of reasons for this. The first is partly historical. Having become well established, with a large number of transmitter networks and many millions of receivers it is not possible to discontinue the service easily. Receivers for AM can also be made very cheaply and easily. Another factor is that the frequency allocations have many advantages in terms of coverage and propagation. However, these frequencies are not suitable for FM in view of the propagation characteristics as the multipath effects resulting from ionospheric propagation would lead to severe distortion. There is also insufficient spectrum available on these frequencies to allow for the high quality wide-band FM transmissions.

The long wave band is not available for broadcasting in all areas of the world. However, the medium wave band is a worldwide allocation. In view of the quality which can be attained this modifies the nature of the

Figure 8.1 *Studio of a local radio station (Courtesy of Thames FM)*

broadcasting. In areas where the stations are competing with FM services the medium wave tends to have a high percentage of speech-based programming. However, this does not mean that music programming is not present, although stations using the medium wave band are finding that listeners are migrating to the high quality FM broadcasts.

A variety of transmitter powers are used. In the UK, for example, the BBC operates a number of national networks. There are also independent national networks. Transmitters for these run many kilowatts and can be heard over large distances. In addition to these there are many local stations. These are only intended to have a relatively small coverage area and the transmitters may only run 100 watts or so. In North America it is normal for transmitters to be reduced in power at night when propagation via the ionosphere is possible.

In recent years the broadcast bands have been standardized so that stations transmit on particular channels. The channel spacing on the medium wave band is 9 kHz in Europe and 10 kHz in the USA. This reduces the amount of interference because the number of annoying heterodynes from stations 1 or 2 kilohertz away from one another are eliminated.

Short wave broadcasting is still very popular and this is proved by the enormous pressure on the allocations which are available. As a result of this further allocations have been given to broadcasters in a recent World Administrative Radio Conference (WARC) where international frequency spectrum allocations are discussed.

On the short wave bands stations are almost exclusively used for international broadcasting. Because of the nature of propagation at these frequencies, it is unlikely that these stations will achieve much coverage within their own country, unless the skip distance falls within the country. As a result short wave stations are used as flagships for their countries. Often they carry propaganda. This was particularly true in the days of the cold war. Stations from the West could be heard giving exactly the opposite views to those from the Eastern bloc. Even today short wave broadcast stations are still used for this purpose. However, many stations seek to give their countries credibility on the international scene by producing good reliable programming and news. The BBC World Service is recognized the world over as possibly the best station. There are also many religious stations on the air including stations like Vatican Radio and HCJB which broadcasts from the Andes.

Sometimes the short wave bands are used for domestic broadcasting. Many of the countries in the tropical areas of the world are large and relatively sparsely populated. This means that it is not possible to obtain adequate

Figure 8.2 *Short wave broadcasts are heard in many parts of the world (Courtesy of the BBC)*

coverage with either medium wave or VHF FM transmissions. In countries like the UK 20 or more stations may be required to give reasonable coverage for a national network. In countries which are larger and less densely populated it is not economically viable to operate as many transmitters. Coupled with this it is just as important to ensure that the population is able to pick up radio broadcasts to keep them aware of the news and national events. Having a radio station also helps to maintain a sense of national identity, a fact which may be a political necessity.

The solution is to use frequencies which enable a greater coverage to be gained. As a result a number of 'tropical bands' are allocated with frequencies up to about 5 MHz. By using the relatively short skip normally present on these bands greater coverage can be achieved. Often transmitters in the medium wave and VHF FM are used for the populated areas while the tropical bands are used to give greater coverage for the outlying areas.

The tropical bands are only used by countries between latitudes of 23°N and 23°S. This area covers part of Africa, Asia, Central and South America.

With the large amount of pressure on the short wave broadcast allocations it is necessary for broadcasters to make the maximum use of the available

Table 8.1 Long, medium and short wave broadcast bands

Long wave	0.150–0.285
Medium wave	0.5265–1.6065
120 metres	2.300–2.495*
90 metres	3.200–3.400*
75 metres	3.900–4.000†
60 metres	4.750–5.060*
49 metres	5.950–6.200
41 metres	7.100–7.300
31 metres	9.500–9.990
25 metres	11.650–12.050
22 metres	13.600–13.800
19 metres	15.100–15.600
16 metres	17.550–17.900
13 metres	21.450–21.850
11 metres	25.670–26.100

All frequencies are in MHz
* Tropical bands only for use in tropical areas
† Only allocated for broadcasting in Europe and Asia

spectrum. Although the channel spacing is only 5 kHz instead of the greater spacing used on the medium wave band spectrum availability is still tight. To improve the spectrum usage and reduce congestion it is planned that broadcasts will start to use single sideband. While it is realized that this cannot occur overnight, it is planned that a switch to this mode will be made a number of years into the twenty-first century. The exact date of the changeover has recently been deferred. When broadcast stations use SSB, they will only transmit one sideband. Also the level of the carrier will be reduced by 6 dB which will enable major savings to be made in the power required for the transmitter.

VHF FM

One of the main disadvantages of AM is the fact that the transmissions are subject to interference and noise. The bandwidths used also mean that the audio frequency response is limited. The VHF FM broadcasts offer much higher quality both in terms of the noise performance and frequency response, with a deviation of ±75 kHz and an overall transmission bandwidth of 200 kHz. The upper audio frequency limit is generally taken as 15 kHz for these transmissions. This is quite adequate for most high quality transmissions.

One of the problems with these high quality VHF FM transmissions is that the increased audio bandwidth means that noise can often be perceived. Even then it is considerably better than that obtained using an AM system. It is particularly noticeable towards the treble end of the audio spectrum, where it can be heard as a background hiss. To overcome this it is possible to increase the level of the treble frequencies at the transmitter. At the receiver they are correspondingly attenuated to restore the balance. This also has the effect of reducing the treble background hiss which is generated in the receiver. The process of increasing the treble signals is called pre-emphasis, and reducing the in the receiver is called de-emphasis. The rate of pre-emphasis and de-emphasis is expressed as a time constant. It is the time constant of the capacitor-resistor network used to give the required level of change. In the UK, Europe and Australia the time constant is 50 μs whereas in North America it is 75 μs.

Stereo

In line with the improved quality available on the VHF FM broadcasts, it is possible to transmit stereo. However, this has to be accomplished in such a way that ordinary mono radios can still receive the transmissions without any degradation in performance.

A stereo signal consists of two channels which can be labelled L and R (Left and Right), one channel for each speaker. The ordinary mono signal

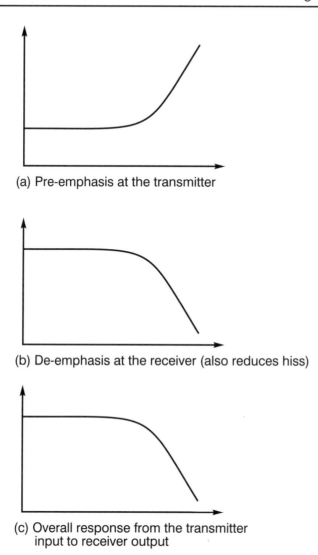

(a) Pre-emphasis at the transmitter

(b) De-emphasis at the receiver (also reduces hiss)

(c) Overall response from the transmitter
input to receiver output

Figure 8.3 *Pre-emphasis and de-emphasis of a signal*

consists of the summation of the two channels, i.e. L + R, and this can be transmitted in the normal way. If a signal containing the difference between the left and right channels, i.e. L − R, is transmitted then it is possible to reconstitute the left and right only signals. By adding the sum and difference signals, i.e. (L + R) + (L − R), gives 2L, i.e. the left signal, and subtracting the two signals, i.e. (L + R) − (L − R), gives 2R, i.e. the right signal. This can be achieved relatively simply by adding and subtracting the two signals electronically. It only remains to find a method of transmitting the stereo difference signal in a way that does not affect any mono receivers.

This is achieved by transmitting the difference signal above the audio range. It is amplitude modulated onto a 38 kHz subcarrier. Both the upper and lower sidebands are retained, but the 38 kHz subcarrier itself is suppressed to give a double sideband signal above the normal audio bandwidth as shown in Figure 8.4. This whole baseband is used to frequency modulate the final radio frequency carrier. This signal is also what is regenerated after the signal is demodulated in the receiver.

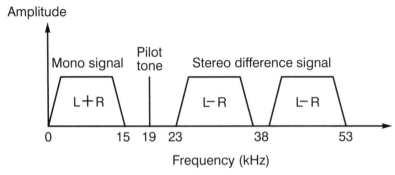

Figure 8.4 *The modulating (baseband) signal for a stereo VHF FM transmission*

To regenerate the 38 kHz subcarrier, a 19 kHz pilot tone is transmitted. The frequency of this is doubled in the receiver to give the required 38 kHz signal to demodulate the double sideband stereo difference signal.

The presence of the pilot tone is also used to detect whether a stereo signal is being transmitted. If it is not present the stereo reconstituting circuitry is turned off. However, when it is present the stereo signal can be reconstituted.

To generate the stereo signal, a system similar to that shown in Figure 8.5 is used. The left and right signals enter the encoder where they are passed through a circuit to add the required pre-emphasis. After this they are passed into a matrix circuit. This adds and subtracts the two signals to provide the L + R and L − R signals. The L + R signal is passed straight into the final summation circuit to be transmitted as the ordinary mono audio. The difference L − R signal is passed into a balanced modulator to give the double sideband suppressed carrier signal centred on 38 kHz. This is passed into the final summation circuit as the stereo difference signal. The other signal entering the balanced modulator is a 38 kHz signal which has been obtained by doubling the frequency of the 19 kHz pilot tone. The pilot tone itself is

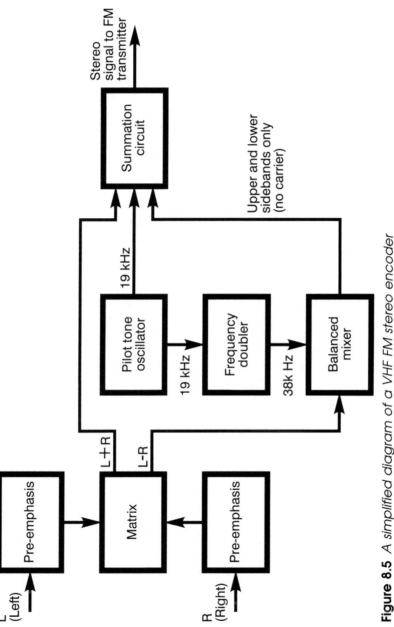

Figure 8.5 *A simplified diagram of a VHF FM stereo encoder*

also passed into the final summation circuit. The final modulating signal consisting of the L + R mono signal, 19 kHz pilot tone, and the L − R difference signal based around 38 kHz is then used to frequency modulate the radio frequency carrier before being transmitted.

Reception of a stereo signal is very much the reverse of the transmission. A mono radio receiving a stereo transmission will only respond to the L + R signal. The other components being above 15 kHz are above the audio range, and in any case they will be suppressed by the de-emphasis circuitry.

A variety of methods are available for decoding the stereo signal once it has been demodulated in the radio. The basic method and circuitry for demodulating an FM signal are described in Chapter 5 (Receivers). This produces the baseband signal consisting of the L + R, L − R, and the pilot tone. These have to be extracted and converted into the two audio signals L and R.

The block diagram of one type of decoder is shown in Figure 8.6. Although this is not the only method which can be used it shows the basic processes which are required. The signal is first separated into its three constituents. The L + R mono signal between 0 and 15 kHz, the pilot tone at 19 kHz, and the stereo difference signal situated between 23 and 53 kHz. First the pilot tone at 19 kHz is doubled in frequency to 38 kHz. It is then fed into a mixer with the stereo difference signal to give the L − R signal at audio frequencies. Once the L + R and L − R signals are available they enter a matrix where they are added and subtracted to regenerate the L and R signals. At this point both signals are amplified separately in the normal way in a stereo amplifier before being converted into sound by loudspeakers or headphones.

Today most stereo radios use an integrated circuit to perform the stereo decoding. Often the pilot tone is extracted and doubled using a phase locked loop. This provides a very easy and efficient method of performing this function without the need for sharp filters.

C-QUAM

With the migration of many listeners from the medium wave bands to the higher quality transmissions on VHF FM many AM broadcasters sought ways of improving the transmissions used on the medium wave bands. One way was to introduce a stereo capability, a feature which is now standard on VHF FM. A number of systems were developed to provide this on AM, but the one which has gained the most widespread acceptance is called C-QUAM (Compatible QUadrature Amplitude Modulation). This system is most widespread in North America, and is used in a number of other countries outside this area, although there is little use within Europe.

In order to see how C-QUAM works the first stage is to look at how a basic quadrature system works. Here two audio channels are placed onto the same radio frequency carrier, by placing the sidebands for one at 90° to the other.

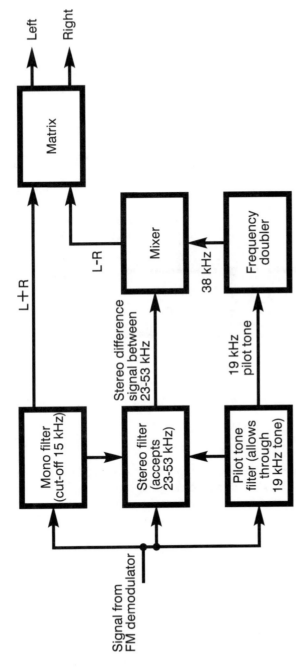

Figure 8.6 *Block diagram of a stereo decoder*

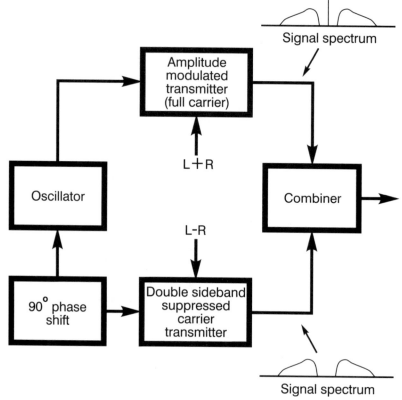

Figure 8.7 *System to generate a basic quadrature AM signal*

This can be achieved relatively simply as shown in Figure 8.7. A standard AM transmitter is used as normal to carry the L + R signal. The sidebands for the difference (L − R) signal are generated by a second transmitter. This one only generates the sidebands as no carrier is required. To obtain the quadrature phasing between the two sets of sidebands a 90° phase shift is added to the master oscillator signal for the second transmitter. Finally the outputs from the two transmitters are added together to give a signal which has a carrier with one set of sidebands carrying the normal mono or sum signal, and then a second set of sidebands carrying the stereo difference signal at 90°.

Figure 8.8 shows a suitable system for demodulating a signal of this nature. When the signal enters the demodulator it is fed straight into two balanced demodulators or product detectors. These are mixers, and are able to detect the audio on the carrier. However, they are also phase sensitive, and

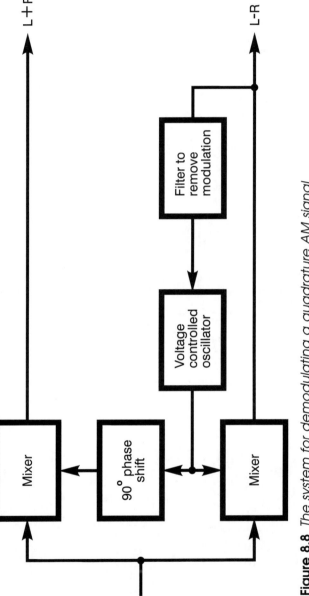

Figure 8.8 *The system for demodulating a quadrature AM signal*

by adjusting the phase of the reference oscillator it is possible to extract one sideband or the other.

This type of quadrature signal works well if it is to be demodulated in this way. However, most mono AM radios detect the signal using a diode envelope detector. These detectors only see the amplitude information and do not take account of any of the phase information carried on the signal. As a result the signal can become very distorted under some conditions. To see how this happens it is necessary to look at two extremes: one when the left and right channels carry are the same; and the other when only the left or the right channel carries no sound. In the first instance when the two channels are the same, the L − R signal is zero because there is no difference, and this makes the second set of sidebands zero, making this transmission case exactly the same as ordinary AM. When only the left or right channel is

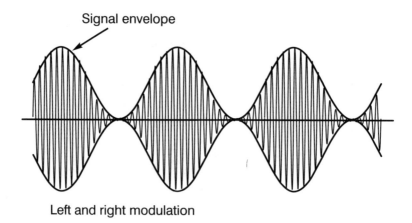

Signal envelope

Left and right modulation

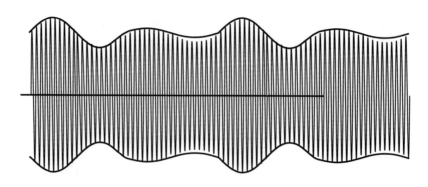

Left or right only

Figure 8.9 *Distortion on a quadrature AM signal*

present the situation is very different. Where both the L + R or sum signal and the L – R or difference signal are present, and both transmitters in Figure 8.9 contribute equally to the transmitted signal making an envelope like that shown in Figure 8.9. If this signal is received on a standard mono receiver it will sound very distorted. Similar levels of distortion will appear when both channels have signals but are different. Although these are extreme cases, it can be seen that some distortion will occur when there are differences between the channels, and this will clearly not be acceptable for listeners with mono receivers.

The system which is finally employed modifies the signal to make it totally compatible with normal AM. The C-QUAM decoder takes account of this when it decodes the stereo signal. In generating a C-QUAM signal, the transmitter is modulated in the standard way by the L + R signal. Circuitry is then used which takes the phase information from the quadrature signal, and this is applied as phase modulation to the master oscillator. In this way a minimum amount of modifications need to be performed on the transmitter to convert it to C-QUAM.

Demodulation of the C-QUAM signal is performed in two stages by circuitry outlined in the block diagram in Figure 8.10. Initially the L + R signal is decoded. Once this has been done the incoming C-QUAM signal is converted into pure quadrature modulation, after which the L – R signal can be extracted as before. When the signal enters the demodulator the amplitude modulation is detected to retrieve the L + R signal. This is used to drive the comparator. The other signal entering the comparator comes from the quadrature modulation detector. This signal would be the L + R signal if pure quadrature modulation was used. As there is a difference between the two this appears at the output of the comparator, and it is used to alter the signal level using the gain modulator so that the correct envelope for pure quadrature modulation is regenerated. Once this has occurred the L – R signal can be regenerated by using the balanced modulators as shown. Having extracted the L + R and L – R signals the two signals are added and subtracted in a matrix circuit to give the L and R signals on their own. These can be amplified in the normal way by a stereo amplifier before being converted into audio energy by headphones or loudspeakers.

RDS

The initials RDS stand for Radio Data System. It is a system which uses inaudible data signals added to a VHF FM transmission to bring a variety of information and automatic tuning facilities to the listener. Initially aimed towards the car radio market RDS is now a standard feature on many hi-fi tuners and other sets. The system is well established in Europe and is used in a number of other countries as well.

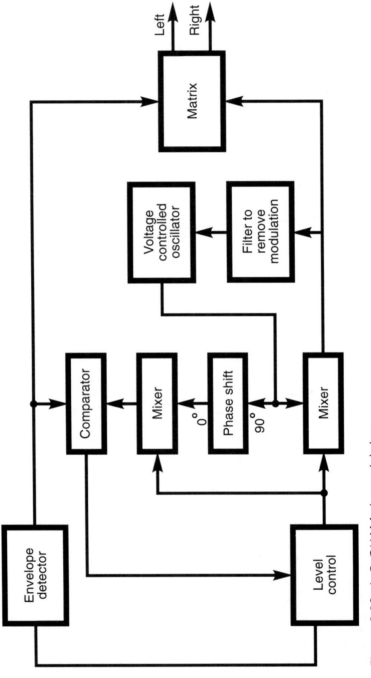

Figure 8.10 *A C-QUAM demodulator*

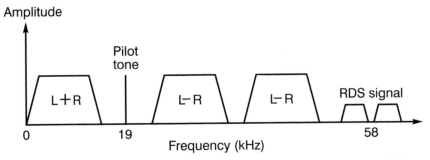

Amplitude

Pilot
tone

L+R

L−R

L−R

RDS signal

0   19

Frequency (kHz)

58

Figure 8.11 *Baseband modulation for an FM signal carrying RDS*

The system operates by adding a data signal to the baseband modulation. It is carried above the stereo difference signal on a 57 kHz subcarrier. This is three times the stereo pilot carrier, as shown in Figure 8.11. The data is modulated onto this subcarrier using a form of modulation called Quadrature Phase Shift Keying (QPSK). This gives good immunity to data errors while still allowing sufficient data to be carried. Being a phase shift keying system it also gives good immunity to interference to the audio signal being carried.

Data is transmitted at a rate of 1187.5 bits per second. This is equal to the RDS subcarrier frequency divided by 48, and by adopting this approach the decoding circuits are able to operate synchronously, reducing the problems of spurious signals.

Data is transmitted in groups consisting of four blocks, each containing a 16 bit information word and a 10 bit check word as shown in Figure 8.12. The large check word is required to ensure the system operates correctly even under conditions of low signal strength or high interference. Once detected it

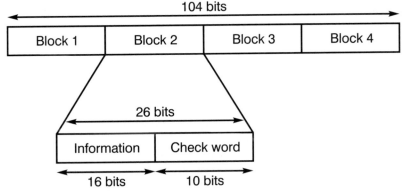

104 bits

| Block 1 | Block 2 | Block 3 | Block 4 |

26 bits

| Information | Check word |

16 bits   10 bits

Figure 8.12 *RDS data structure*

also allows the receiver to correct errors, thereby making the system far more resilient.

Once generated the data is encoded onto the subcarrier in a differential format. When the data level is at a logical '0' the output remains unchanged. When a logical '1' appears the output changes state. The baseband signal has to have its spectrum carefully limited. This is done to ensure that no crosstalk between any of the decoders takes place. This is achieved by encoding each bit as a biphase signal and also by passing the coded signal through a filter.

RDS provides a variety of facilities not previously available with VHF FM broadcasts. The most widely publicized is the travel service which is of particular use for car radios. Stations providing this service transmit the TP code to identify travel messages. When this occurs the volume may be increased, a tape paused, etc. so that the travel message can be heard. After the message is complete the receiver is returned to its normal state.

RDS also enables tuning facilities to be improved. Again this is very useful for car radios. As the car moves from the service area of one transmitter to the next it can automatically tune to the strongest transmitter. The set accomplishes this by looking at the alternative frequency (AF) list and checks that signals on these channels have the correct programme identity (PI) code. When radios fitted with RDS store the frequency of a station, they also store the programme identification (PI) code along side it. In this way if the receiver is located outside the coverage area for the given transmitter it can still locate the correct network.

While the system operates in this manner for national networks some changes have to be made for local stations. Here the radio will tune to the strongest station of the same type when moving out of the service area of the first station.

The PI code is structured, consisting of four characters. The first indicates the country of origin, and it is 'C' for the UK. The second indicates the type of coverage. The figure '2' indicates a national station, and the final two letters are used for the programme reference. As an example C203 is the code for the BBC Radio 3 network and C311 is used for the BBC local radio station GLR.

It can be imagined that when moving from the service area of one transmitter to the next there may be a small gap as the receiver checks for the strongest signal, and that it has the correct PI. This can be improved by employing a second front end section in the set to check the alternative frequencies. Some sets incorporate this and the change from one transmitter to the next is virtually unnoticeable.

Another facility which is becoming increasingly useful is the Programme Service Name (PS). This enables the receiver to display the station name, a facility which is very useful with the large variety of stations on the air today.

Table 8.2 RDS terminology

AF   Alternative Frequencies
  A list of the frequencies used by a station in adjacent coverage areas.

CT   Clock Time and Date
  Data containing time and date information. This enables the clock to display the correct time and adjust between changes in winter and summer time without the need for manual setting.

DI   Decoder Information
  This signal allows for miscellaneous functions to be controlled in the radio.

EON   Enhanced Other Networks
  Information which is transmitted giving the radio a cross-reference to other stations when travel information is being transmitted.

MS   Music/Speech
  A data flag which allows for the relative levels of speech and music to be altered.

PI   Programme Identification
  This is a station code used in conjunction with the alternative frequency signal to locate suitable alternative transmitters for a given programme. Each service is allocated its own unique identifier.

PIN   Programme Identification Number
  This signal identifies a given programme, and allows for a radio or recorder to be turned on when the identification number is recognized.

PTY   Programme Type Selection
  This signal identifies one of 15 types of programme, allowing selection of the type of listening rather than by station.

PS   Programme Service Name
  A signal which carries the station name and allows this to be displayed by the receiver.

RT   Radio Text
  This allows information about the programme to be displayed by the radio.

TDC   Transparent Data Channel
  This allows data to be downloaded over the radio.

TP/TA   Travel Service
  These signals enable the travel information to be heard, regardless of the choice of listening.

A further facility which was not available at the launch of RDS is called Enhanced Other Networks (EON). This allows the set to listen to a national network, but to be interrupted by travel news on a local station. This enables the set to be sensitive only to relatively local traffic problems even though the set is tuned to a national network. To achieve this facility requires a large amount of coordination between the different stations. To implement this feature the BBC have a central computer specifically for this. When a local station is about to transmit a traffic message this fact is communicated to this computer. It directs the relevant transmitters in the national network to indicate this fact, enabling the receiver to retune to the relevant local radio station.

To be able to receive RDS data three circuit blocks are required. The first is a 57 kHz filter which is used to isolate the signals from the baseband audio and the stereo difference signals. Next this must be fed into a demodulator. This removes the data from the 57 kHz subcarrier. Once the data is available it needs to be processed for use by the receiver. As most car radio sets these days incorporate a processor, it can also be used to perform this function.

RDS has proved to be very popular for VHF FM. A similar scheme is also being investigated for short wave transmitters where similar problems of not being able to identify stations and not knowing the alternative frequencies are experienced. Called AMDS (Amplitude Modulation Data System) it has many similarities to VHF FM RDS, although the data is carried by phase modulating the carrier.

Digital audio broadcasting

Most current broadcast techniques use analogue techniques to carry the audio on the carrier. However, with the move to the use of digital techniques because of the improved performance a digital system is now available in a number of countries. Called Digital Audio Broadcasting (DAB) it gives many improvements and new features over the existing AM and FM systems in use.

DAB gives major improvements over FM, particularly for those in automobiles. The existing FM system suffers, particularly when signal strengths are low, and the signal is being received by a number of different paths. Reflections off buildings, hills and other objects cause distortion of the signal, and in addition to this the tuning has to be altered when passing from the coverage area of one transmitter to the next. With the increase in the number of facilities being expected, new data capabilities are also required. All of these have been incorporated into DAB together with an improved ease of tuning.

The problem of reflections is at the core of determining how the new system will work. Although digital audio is widely used in television broadcasts, these have the advantage that direction antennas are used which

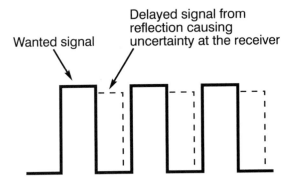

Figure 8.13 *Data corruption caused by reflections*

will greatly limit the effect of any reflections. For car systems which use an omnidirectional antenna, any system must be resilient to reflections, without suffering any degradation. The system is also designed to make tuning easier. On the medium wave or VHF FM bands, there are many stations which can be heard, often making it difficult to locate the wanted station. DAB aims to simplify this to a very large degree.

The main problem with a digital system is to ensure that reflections do not corrupt the data being sent, as shown in Figure 8.13. For satisfactory operation the system must be able to cope with significant delays. This reduces the data rate well below that required for a system using normal linear encoding like that used for a CD. For these discs the data rate is about 1.4 Mbits per second.

To enable the data to be successfully transmitted, two data rate reduction systems are employed. The first is an audio processing system, and the second is a totally different form of modulation called Coded Orthogonal Frequency Division Multiplex (COFDM).

The audio processing system relies upon the fact that the ear can only detect certain sounds. Below a given level it does not hear them. This threshold of hearing varies with frequency, as shown in Figure 8.14. The presence of sounds also varies the threshold. A strong signal in one portion of the audio spectrum will mask weaker ones close to it as shown. By detecting the threshold of hearing and not encoding signals below that level the data rate can be significantly reduced, in this system by a factor of six when compared to a linearly encoded system.

The modulation system uses a large number of low data rate carriers packed closely together. This is achieved by using frequency division mutliplexing, and overlapping the signals slightly. However, by altering the phase so that they are orthogonal to one another the individual carriers can be demodulated without any interference from one another.

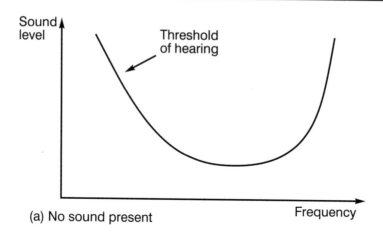

(a) No sound present   Frequency

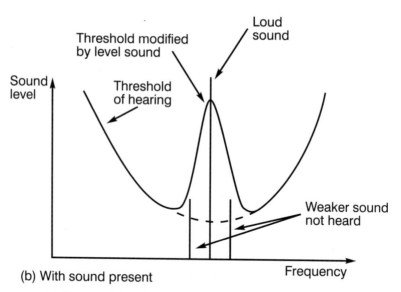

(b) With sound present   Frequency

Figure 8.14 *Threshold of hearing of the ear*

A total of just over 1500 carriers occupying about 1.5 MHz of bandwidth is used to carry the data, and they have a spectrum like that shown in Figure 8.15. A guard band of 250 kHz is placed between adjacent channels to prevent interference between different transmissions. This allows data at a rate of just over 2.4 Mbits per second to be carried giving the capability of up to six high quality stereo transmissions to be carried or 20 or so low quality mono services. It is possible to alter the data rate according to the

Amplitude

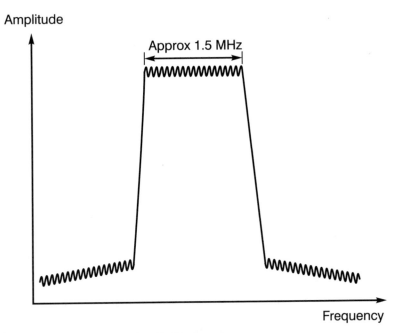

Figure 8.15 *Spectrum of a DAB signal*

requirements of the programme so that maximum use is made of the available capacity. Additional data services are also available.

The DAB transmissions are very efficient in their use of spectrum. In view of their resilience to interference it means that adjacent transmitters do not need to use different frequencies as is the case for normal AM and FM services. Instead the same frequency is reused as required to give continuous coverage in what is termed a single frequency network (SFN). In the UK, a single national network for FM occupies 2.2 MHz. A DAB transmission occupying only 1.5 MHz can carry several audio services as well as having an additional data capability.

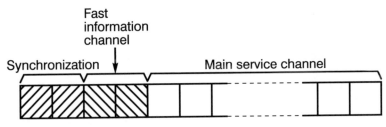

Figure 8.16 *DAB data frame structure*

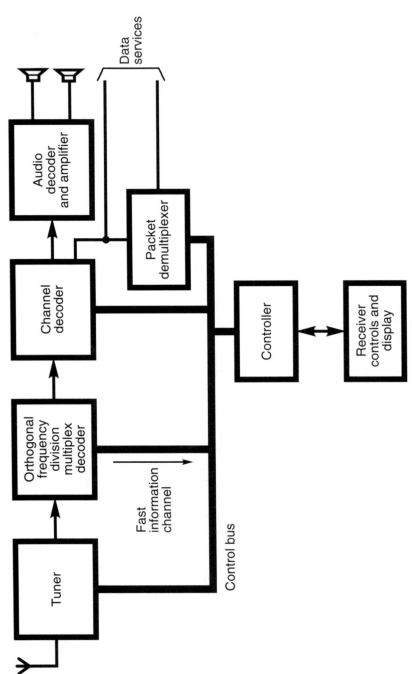

Figure 8.17 *Block diagram of a DAB receiver*

To enable the receiver to be able to decode the data being transmitted it must have a defined format. This is shown in Figure 8.16. Each transmission frame starts with a null signal to give approximate synchronization. This is followed by a phase reference for the demodulation process. After these references the next symbols of data are reserved for the Fast Information Channel (FIC) used for data. The remaining frames are then used for the Main Service Channel (MSC). Each programme or service within the MSC section of the frame is allocated its own time for decoding purposes. The total length of the frame for terrestrial transmissions which are being made at the moment is 96 ms, although additional standards exist for 48 or 24 ms. These are intended for use on other frequencies and for satellite broadcasts which are planned for the future.

Receivers for DAB naturally require totally different decoding circuitry than that used for more typical AM or FM receivers. They make heavy use of digital signal processing techniques to reconstruct and process the data. A typical receiver block diagram is shown in Figure 8.17. Circuitry up to the intermediate frequency stages is similar to that of a normal radio receiver, using the superhet principle. Even so the IF stage must be sufficiently wide to accommodate the 1.5 MHz signal. After the IF stages the similarities end because the signal enters an orthogonal frequency division multiplex decoder. This reconstitutes the raw data which is further processed as shown to produce the audio for amplification, or data for display or other uses, including being linked into PCs etc.

DAB is still in its early stages of use, but the advantages it holds over AM and VHF FM transmissions will ensure that it will become firmly established in the near future.

9 Satellites

Satellites have become an accepted part of today's technology. They are used for a variety of purposes from providing links for long distance tele-communications, to direct broadcast television and radio, gathering weather and geological information, as well as a host of other facilities. To reach where we are today has taken a tremendous amount of development and investment since the idea of communications satellites was first proposed by Arthur C. Clarke in an article in *Wireless World* in 1945. Since then the electronics to enable all the functions in the satellite to be performed has been developed along with the launch vehicles capable of putting these satellites, often as heavy as a ton, into space. Both of these elements of the development have required a large investment. At the time of Arthur C. Clarke's article the most advanced rockets were the V2s which had been developed by the Germans, and these had never been intended to launch satellites into outer space. For the satellites themselves methods of powering them had to be developed along with control systems to enable them to be finely positioned. The satellites also had to be made very reliable as they are not able to be serviced once they have been launched. Nowadays all these problems have been mastered and satellites are capable of providing a long, reliable service life.

Satellite orbits

There are a variety of different orbits which can be adopted by satellites. The one which is chosen will depend on factors including its function, and the area it is to serve. In some instances the orbit may be as low as 100 miles, whereas others may be over 22 000 miles high.

As satellites orbit the earth they are pulled back in by the force of the gravitational field. If they did not have any motion of their own they would fall back to earth, burning up in the upper reaches of the atmosphere. Instead the motion of the satellite rotating around the earth has a force associated with it pushing it away from the earth. For any given orbit there is a speed for which these two forces balance one another and the satellite remains in the same orbit, neither gaining height nor losing it.

Obviously the lower the orbit, the stronger the gravitational pull, and this means that the satellite must orbit the earth faster to counteract this pull. Further away the gravitational field is less and the satellite velocities are correspondingly less. For a very low orbit of around 100 miles a velocity of about 17 500 miles per hour is needed and this means that the satellite will orbit the earth in about 90 minutes. At an altitude of 22 000 miles a velocity of just under 7000 miles per hour is needed giving an orbit time of about 24 hours.

A satellite can orbit the earth in one of two basic types of orbit. The most obvious is a circular orbit where the distance from the earth remains the same at all times. A second type of orbit is an elliptical one. Both types of orbit are shown in Figure 9.1, where the main characteristics are shown.

When a satellite orbits the earth, either in a circular or elliptical orbit it forms a plane. This passes through the centre of gravity of the earth or the

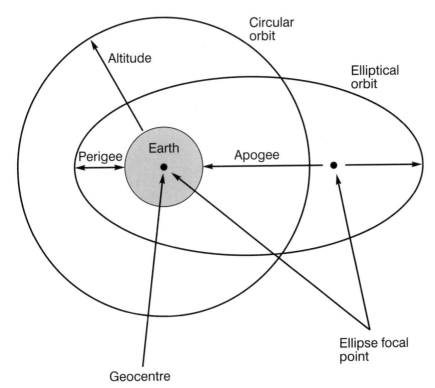

Figure 9.1 *Satellite orbits*

geocentre. The rotation around the earth is also categorized. It may be in the same direction as the earth's rotation when it is said to be posigrade, or it may be in the opposite direction when it is retrograde.

For many orbit calculations it is necessary to consider the height of the satellite above the geocentre. This is the height above the earth plus the radius of the earth. This is generally taken to be 3960 miles or 6370 km.

Velocity is another important factor, as already seen. For a circular orbit it is always the same. However, in the case of an elliptical one this is not the case as the speed changes dependent upon the position in the orbit. It reaches a maximum when it is closest to the earth and it has to combat the greatest gravitational pull, and it is at its lowest speed when it is furthest away.

For an elliptical orbit the centre of the earth forms one of the focal points of the ellipse, as shown in Figure 9.1. It can also be seen that there are points where the satellite is furthest from the earth, and closest to it. These are important and they are called the apogee (furthest away), and perigee (closest). Generally the altitudes at the apogee and perigee are calculated from the geocentre.

A satellite may orbit around the earth in different planes. The angle of inclination of a satellite orbit is shown in Figure 9.2. It is the angle between a line perpendicular to the plane of the orbit and a line passing through the poles. This means that an orbit directly above the equator will have an inclination of 0° (or 180°), and one passing over the poles will have an angle of 90°. Those orbits above the equator are generally called equatorial obits, while those above the poles are called polar orbits.

Another important factor about the position of a satellite is its angle of elevation. This is important because the earth station will only be able to maintain contact with the satellite when it is visible as only direct or line of

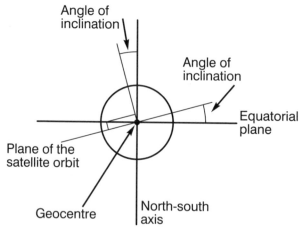

Figure 9.2 *Angle of inclination of an orbit*

sight communications are possible with the satellite. The angle of elevation is the angle at which the satellite appears above the horizontal. If the angle is too small then signals may be obstructed by nearby objects if the antenna is not very high. For those antennas which have an unobstructed view there are still problems with small angles of elevation. The reason is that signals have to travel through more of the earth's atmosphere and are subjected to higher levels of attenuation as a result. An angle of 5° is generally accepted as the minimum angle for satisfactory operation.

In order that a satellite can be used for communications purposes the ground station must be able to follow it to receive its signal, and transmit back to it. Communications will naturally only be possible when it is visible, and dependent upon the orbit it may only be visible for a short period of time. To ensure that communication is possible for the maximum amount of time there are a number of options which can be employed. The first is to use an elliptical orbit where the apogee is above the planned earth station so that the satellite remains visible for the maximum amount of time. Another option is to launch a number of satellites with the same orbit so that when one disappears from view, and communications are lost, another one appears. Generally three satellites are required to maintain almost uninterrupted communication. However, the handover from one satellite to the next introduces additional complexity into the system, as well as having a requirement for three satellites.

The most popular option is to use a satellite in what is called geostationary orbit. Using this orbit the satellite rotates in the same direction as the rotation of the earth and has a 24 hour period. In this way it revolves at the same angular velocity as the earth and in the same direction. As a result it remains in the same position relative to the earth. For the relative forces to balance and the satellite to remain in a geostationary orbit it must be above the equator and have an altitude of 22 300 miles or 35 860 km. Geostationary orbits are very popular because once the earth station is set onto the satellite it can remain in the same position, and no tracking is normally necessary. This considerably simplifies the design and construction of the antenna. For direct broadcast satellites it means that people with dishes outside the home do not need to adjust them once they have been directed towards the satellite.

Satellites

The requirements for satellites are very stringent, both in terms of their design and construction. They must be capable of operating in extreme conditions while still maintaining the highest standards of reliability because they cannot be retrieved for maintenance or repair. Satellites also contain a number of systems used for what is called station keeping.

These systems are used to maintain the satellite in the correct orbit and position. The satellite will tend to drift away from its correct position over a period of time. To correct this small thrusters are used. Often they consist of canisters of a gas which when released with a catalyst gives a form of rocket propulsion to move the satellite back on station. Often the service life of a satellite is determined by the amount of fuel for repositioning the satellite rather than the reliability of the electronics.

The other problem with a satellite is that its attitude will change. This is of great importance because directive antennas or cameras are often used, and the satellite needs to be orientated in the correct direction for them. The basic method of gaining the correct orientation is to use the thrusters. However, the attitude will change comparatively quickly. The most common method to overcome this is to use the gyroscopic effect. Sometimes a large flywheel may be made to spin inside the satellite. This can be inefficient in its use of the weight of the satellite. To overcome this other cylindrical satellites actually rotate a portion of the body, often an inner cylindrical section so that the antennas mounted on the outer section do not revolve.

Electrical power is also required by the satellite for its electronic circuitry and other electrical systems. This is supplied by the large arrays of photo or solar cells. Some cylindrical satellites have them positioned around the outer area on the cylinder so that some part of the body is always exposed to sunlight. Others have large extending panels which are orientated to collect the maximum amount of light. Today these panels are capable of producing the many kilowatts of power required for the high power output stages used in many transponders.

The satellite also needs reserve batteries which can be charged from the main solar cells. These are required for periods when the satellite is not in sunlight and should be sufficient to power the satellite for the full period of darkness.

Satellites must also be designed and manufactured to withstand the harsh environment encountered in outer space. Extremes of temperature are encountered. The surfaces exposed to the sun are heated by solar radiation and will rise to very high temperatures, whereas the other side which is not heated will be exceedingly cold. Only conduction will give any heating effect under these circumstances.

There are a number of other effects which must be considered. Solar radiation itself has an effect on some materials, causing them to degrade. Meteorites are another problem. Very small ones cause the surfaces to be eroded slightly, but larger ones may penetrate the body of the satellite causing significant damage. To overcome this satellites are protected by specially designed outer layers. These consist of sheets of metal which are slightly separated giving a cushioning effect when any meteorites impact on the satellite. Cosmic particles also degrade the performance of satellites. Particularly during solar flares the increase in solar particle flow can degrade solar cells, reducing their efficiency.

Placing a satellite in orbit

In view of the colossal amount of energy required to place a satellite in orbit it is necessary to ensure that the energy is used in the most effective way. This ensures that the amount of fuel required is kept to a minimum; an important factor on its own because the fuel itself has to be transported until it is used.

Many satellites are placed into geostationary orbit, and one common method of achieving this is based on the Hohmann transfer principle. This is the method used when the *Shuttle* launches satellites into orbit. Using this system the satellite is placed into a low earth orbit with an altitude of around 180 miles. Once in the correct position in this orbit rockets are fired to put the

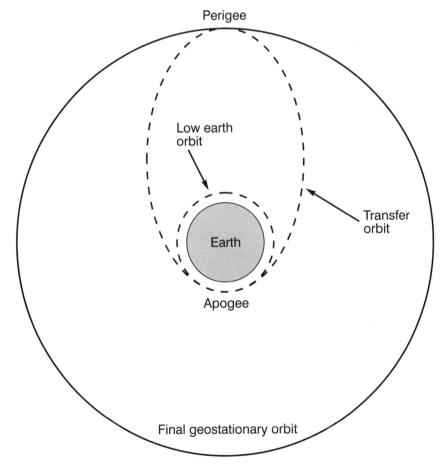

Figure 9.3 *Use of a transfer orbit to place a satellite in geostationary orbit*

satellite into an elliptical orbit with the perigee at the low earth orbit and the apogee at the geostationary orbit as shown in Figure 9.3. When the satellite reaches the final altitude the rocket or booster is again fired to retain it in the geostationary orbit with the correct velocity.

Alternatively when launch vehicles like *Ariane* are used the satellite is launched directly into the elliptical transfer orbit. Again when the satellite is at the required altitude the rockets are fired to transfer it into the required orbit with the correct velocity.

These are the two main methods of placing satellites into orbit. Naturally it would be possible to place a satellite directly into geostationary orbit, but this would take more energy and would not be feasible.

Path calculations

Communication with satellites occurs via a direct line of sight path. As a result it is possible to calculate the signal levels which will be received by the satellite or the earth station assuming that figures like the transmitter power, antenna gains and receiver performance are known. To ensure that the signal levels are kept within working limits without over-designing a system which would cost extra money, link budgets are always calculated.

The major calculation which is performed when investigating the link budget is the path loss. As the signal travels along a direct line of sight the signal attenuation which is experienced is the free space path loss. This is expressed in decibels:

$$\text{Loss (dB)} = 22 + 20 \log_{10} (R/\lambda)$$

where R is the range or distance

λ is the wavelength expressed in the same units as the distance

From this it can be seen that the loss increases with frequency, but it should be remembered that at higher frequencies it is possible to use higher gain antennas and this compensates for this. This means that for a satellite in geostationary orbit the loss will be between 195 and 213 dB assuming operating frequencies between 4 and 30 GHz. Although the altitude of the satellite is 22 300 miles, the actual distance will often be greater than this because the earth station will not be directly below the satellite.

Communications satellites

Satellites fulfil a number of roles. They can be used for communications, direct broadcasting, weather monitoring, navigation and a number of other uses. Of these the most widespread is possibly their use in communications. Every day many thousands of international telephone calls are made. These are usually carried via satellites. In the early days of telephones, calls were

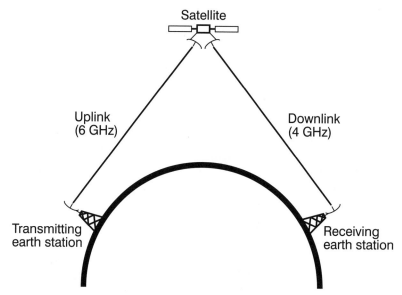

Figure 9.4 *Using a satellite for long distance communications*

routed via undersea cables. These were expensive to lay, and were difficult to maintain. They also had a relatively narrow bandwidth and this meant the number of calls they could carry was limited.

For use in communications like this the satellite acts a repeater. Its height above the earth means that signals can be transmitted over distances which are very much greater than the line of sight as shown in Figure 9.4. An earth station transmits the signal, normally consisting of many telephone calls up to the satellite. This is called the uplink and may be typically on a frequency of 6 GHz. The satellite receives the signal and retransmits the signal. This may typically be on a frequency of 4 GHz on what is termed the downlink. This has to be on a different frequency to avoid interference between the two signals. If the transmission took place on the same frequency as the received signal, then the transmitted signal would overload the receiver in the satellite, preventing reception on the uplink. The use of satellites is very widespread and a wide variety of frequencies are used. While the figures included here are typical, many other frequencies are also used.

The circuitry in the satellite which acts as the receiver, frequency changer, and transmitter is called a transponder. As shown in Figure 9.5, this basically consists of a low noise amplifier, a frequency changer consisting of a mixer and local oscillator, and then a high power amplifier. Figures used here are the same as those mentioned earlier, and are only given as an example. The signal is received and amplified to a suitable level. It is then applied to the

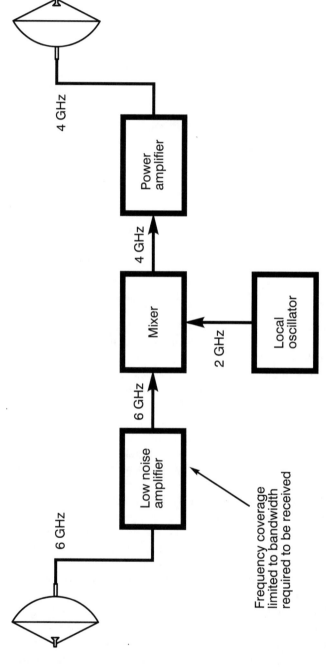

6 GHz

Low noise
amplifier

Frequency coverage
limited to bandwidth
required to be received

6 GHz

Mixer

2 GHz

Local
oscillator

4 GHz

Power
amplifier

4 GHz

Figure 9.5 *Block diagram of a basic transponder*

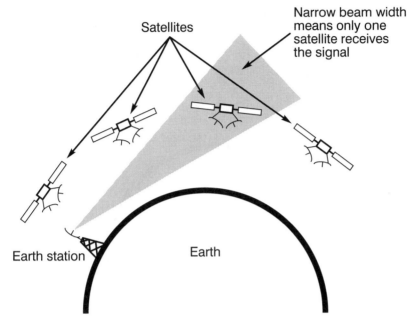

Narrow beam width means only one satellite receives the signal

Satellites

Earth station

Earth

Figure 9.6 *Separating satellites by position*

mixer. This uses the superhet principle to change the signal to the new frequency. After this it is amplified to a suitable level for retransmission.

Signals transmitted to satellites usually consist of a large number of signals multiplexed onto a main transmission. In this way one transmission from the ground can carry a large number of telephone circuits or even a number of television signals. This approach is operationally far more effective than having a large number of individual transmitters.

Obviously one satellite will be unable to carry all the traffic across the Atlantic. Further capacity can be achieved using several satellites on different bands, or by physically separating them from one another. In this way the beamwidth of the antenna can be used to distinguish between different satellites. Normally antennas with very high gains are used, and these have very narrow beamwidths, allowing satellites to be separated by just a degree or so.

Navigational satellites

One of the most recent developments in satellite applications is in the Global Positioning System or GPS. Although initially conceived as a military system, it is being increasingly used for commercial applications. Part of the appeal of the system is that it gives worldwide coverage with accuracies of

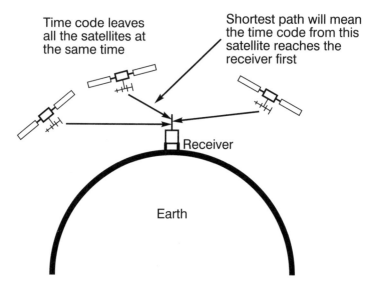

Time code leaves
all the satellites at
the same time

Shortest path will mean
the time code from this
satellite reaches the
receiver first

Receiver

Earth

Figure 9.7 *Operation of GPS*

around 100 metres for the civilian accessible section of the system, and 40 metres for the military section. However, the military system is being made open shortly.

The system consists of 24 satellites placed in a low earth orbit. These transmit on low microwave frequencies and the receivers pick up the transmissions. At least four should be received at once to give good position fixes, and from the information received the receiver is able to compute the latitude, longitude, and altitude.

Each satellite is held in a very accurate orbit. On board the satellite contains an atomic clock to give a highly accurate time source. This time information is transmitted with the signal from each satellite. From this it is possible for the receiver to calculate the different path lengths from the satellites. The time for the signal to travel from the satellite to the earth is proportional to the path length, and since the velocity of the signal is known it is possible to calculate the path length. This information is computed by the processor in the receiver to give the position on the earth.

Other uses

Satellites are used for a variety of other purposes. They can be used for a variety of observation tasks. One of the more familiar of these applications is in weather forecasting where they can take a bird's eye view of the weather formations from high above the earth. This gives weather forecasters a much better view of how the weather formations are developing and moving.

They can also be used in geological surveying. They can observe most of the earth's surface showing the state of crops, forests and the like. Often they can detect the onset or spread of plant diseases in a way that would not be possible using any other means. They can similarly be used in geological surveying where it is possible to detect mineral reserves from the formations seen from outer space.

10 Personal communications

Cellular telephones have brought about a revolution in personal telecommunications over the past few years. Their introduction has enabled people to be contacted almost anywhere. Previously it was necessary to be within reach of a fixed telephone wired into the telephone network. A few mobile phones did exist, but they had limited range and were only available in a few cities. To be able to achieve the flexibility and coverage expected today a totally new system was needed. Systems have evolved in many countries in the world, and although there are differences between countries they use the same basic principles. For example, the system adopted in the UK for analogue systems is TACS (Total Access Communication System), whereas in the USA, AMPS (Advanced Mobile Phone System) has been adopted. For some types of network common standards have been set up allowing users to connect to the systems in other countries. These are digital systems which are increasing in use as they give a number of advantages over their older analogue counterparts.

Basic system

With the vast numbers of people using the cellular phone systems, very efficient use must be made of the available spectrum. With the large numbers of people using cellular phones it is not feasible to allocate each user a different frequency. In addition to this the system must be able to communicate with low powered car phones and portable handsets as it is clearly impracticable to utilize high powered transmitters from the point of view of battery consumption and size. This means that a large number of base stations are required to cover the country.

In view of the fact that low powered handsets are required, this means that only relatively short ranges can be achieved. As a result it is possible to reuse frequencies provided that the phones or base stations are outside the range of each other. In fact in any given country the same frequencies are reused many times over, and if it were not for this fact the cell phone system would not be viable.

To exploit this fact a system of small cells is built up, each one covering a specific area. However, it is not possible to prevent the transmissions from one cell stopping right at the edge of its service area. Instead they will be heard a certain way into the next one. To ensure that transmissions in adjacent cells do not interfere with one another they are allocated different channels. Often a seven cell system like that shown in Figure 10.1 is used. However four, 12 and even 21 cell clusters can be used on some occasions dependent upon the conditions required.

In order that many subscribers can be serviced within each cell there are a number of channels which can be used. These are split into two types: those for transmitting to the base station from the mobile phone, and those for transmitting by the base station to the mobile station. These are split into two bands and arranged so that there is a sufficient bandwidth between the transmitted and received signals. This is required because the same set has to handle signals in both directions at the same time, and interference should not

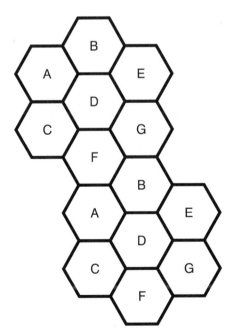

Figure 10.1 *A typical cell system*

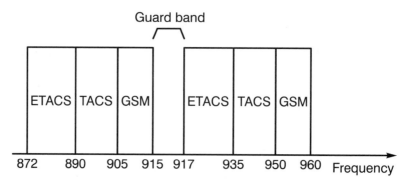

Figure 10.2 *Allocated frequency spectrum for UK mobile phones*

be caused by the transmitted signal to the received one. Particularly the receiver must not be overloaded by the nearby strong transmitted signal.

To give a view of the way in which the channels are split, in the UK the spectrum is allocated as shown in Figure 10.2. There are 300 channels allocated to each of the two operators in the basic TACS system. Of these, 21 are used as control channels, and there are two guard channels between the two operators. This gives 277 channels for carrying the speech information. This gives about 40 channels per cell assuming there are seven cells in each cluster.

An additional 320 channels have been allocated to each operator to allow for the high usage in many areas. These extended channels (Extended TACS) are only accessible by the more recent phones, and therefore when a base station detects an ETACS compatible phone it will allocate it a channel in the ETACS band, allowing those phones which cannot access these channels to access the standard TACS channels. Only when no ETACS channels are free will the base station allocate a TACS channel to an ETACS compatible phone. It will also be noticed there is a considerable spacing between any one pair of channels. This enables sufficient isolation to be obtained between the transmit and receive channels.

The TACS system is used in several countries in Europe including the UK, as well as a number of other countries. However, the AMPS system is more widespread being used in the USA, Canada, and many other countries around the globe. There are many similarities between AMPS and TACS in terms of the basic concept. The basic differences are in the frequencies used, channels spacing and other specifications. The NMT system is also widely used, especially in Northern Europe, but is also found in many other countries around the world.

The planning of the cells is important, and not all cells may be the same size. There are two main reasons for this. The first is that the range is affected by the terrain. In a city, tall buildings may give rise to dead areas behind them. To overcome this many small cells may be needed with small service areas having a radius of about 1 km. The transmitters are likely to be relatively low powered to ensure they do not cause interference with cells further away in directions where they are not shielded. Output powers of about a watt are not unusual in circumstances like these. In rural areas where there are fewer obstructions a single high powered station may cover a much larger area, possibly having a service with a radius of up to 10 km.

The other reason for having smaller cells in cities is that the number of users is generally much higher. In order to ensure that there are sufficient channels to service all the users, the cells must be smaller to reduce the number of users in each cell. Conversely in rural areas there are generally fewer users and consequently fewer cells are needed. The only exception to this is that along major highways usage rises, and these areas must have sufficient cells to ensure that they do not become overloaded.

Base stations

These stations are set up in a variety of places. In towns and cities the characteristic antennas are often seen on the top of buildings, whereas in rural areas separate masts are used. It is important that the height and location of the antennas are carefully chosen to ensure the required coverage is achieved. If the antenna is too low or in a poor location, there will be insufficient coverage. Conversely if the antenna is too high or in a very good location, then the signal will be heard well beyond the boundaries of the cell. This may result in interference with another cell using the same frequencies.

The performance of the equipment in the base station must make up for many of the deficiencies in the mobile set. The base station will often have an output power of around 10 watts. The receiver must be sensitive, the site must be free from high levels of noise which might mask out the wanted signals, and the antennas must have gain to enable the cell signal to be heard throughout the cell, and to enable it to receive signals in the cell. The mobile or portable set on the other hand is likely to have a limited power, especially in the case of a portable hand-held set, the antenna will be small and not offer a good performance, and the location may be very poor as mobile phones are often used inside buildings or in other poor radio locations.

Mobile phones

The mobile phone is a very complicated unit containing a receiver, transmitter and all the logic and control circuitry required to maintain the call. Both transmitter and receiver must be capable of being able to tune to any of the channels allocated, and the control logic must be able to control the operator interface, i.e. the keypad and display, as well as controlling the data sent to and from the base station.

There are three main categories for mobile phones: mobile, transportable and hand portable. Today most of the phones fall into the hand-portable category, many fitting easily into a top pocket. These hand-portable sets only have a limited output power. Generally this is a maximum of about 0.6 watts effective radiated power. This means that in some areas communications may be marginal requiring the user to move to the best position. Mobile and transportable sets have a higher output and will offer more reliable communications at the expense of convenience.

Mobile phones also include a system for reducing the transmitter power. Known as an automatic power control (APC) it reduces the output power of the transmitter when the signal strength is high. The receiver picks up control signals which are routed through to the transmitter to alter the transmitter power as required. By reducing the output power in this way the battery life can be maximized as the transmitter takes the most current. The other advantage is that reducing transmitter powers wherever possible helps to minimize the levels of interference.

There is an enormous variety of phones which are available from a number of different manufacturers. Despite this there are a number of common features. Most phones employ a double conversion superhet to enable them to achieve the required levels of performance. Many have a first IF around 45 MHz to give sufficient reject of image signals and also take the image response out of the mobile phone band. In this way interference from nearby phones is reduced. The second IF is often either 10.7 MHz or 455 kHz and contains a crystal filter to give the required level of adjacent channel rejection.

The local oscillator signal is generated using a frequency synthesizer. In this way the internal processing and logic in the phone can accurately and easily control the transmitter and receiver frequencies.

Batteries for phones are becoming increasingly important. All use packs of rechargeable cells. The older ones still use nickel cadmium (NiCad) packs, but many of the new handsets use nickel metal hydride (NiMH) ones. The new NiMH cells can store more charge in a given volume or weight and can therefore last longer. However, they do need to be charged correctly as they are more easily damaged than their NiCad counterparts. New battery developments mean that new types of battery are appearing. Lithium ion (Li-ion) cells are likely to be used increasingly in the coming years.

In order to enable the mobile phone to be able to communicate in both directions at the same time while using just one antenna, an item called an isolator is required. This consists of two very sharp band-pass filters, one to let the transmitter power through to the antenna in the transmit band, and the other to allow signals from the antenna in the receive and through to the receiver input. In this way the signal from the transmitter in the phone which enters the receiver is minimized.

System control and call routeing

For the cellular system to operate satisfactorily the base stations must be able to route calls through to the ordinary telephone network as well as contacting other mobile phones. Additionally they must be able to control the phones within the cell and pass them on to the next cell when they move out of the service area into the next. This all requires a high degree of control.

To enable all the switching and control functions to take place the base station is linked to a mobile telephone exchange (MTX). The MTX sets up and connects all the calls, routeing them to other mobile phones or to the normal telephone system as shown in Figure 10.3. Sometimes these exchanges are referred to as mobile switching centres (MSCs). It also stores information about the phones including the identification numbers and special features which the subscriber may require.

The MTX also monitors the signals from the mobile phones. When they are in a given cell the signal is monitored and they are serviced by the cell

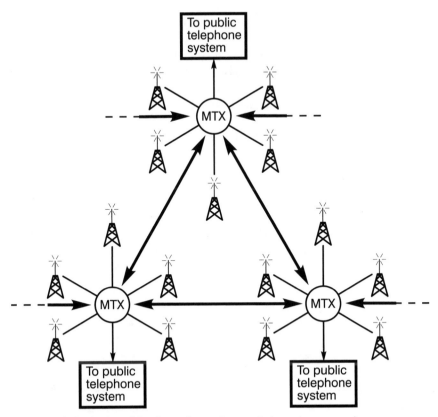

Figure 10.3 *A simplified version of a cellular phone system*

station giving the best signal. As the phone moves out of the service area of a cell the MTX automatically switches the call through to the phone via the next cell. A new channel is allocated for the new cell and the call continues to be routed to the correct destination.

Switching from one cell to the next takes only a small fraction of a second is virtually unnoticed by the user. However, it ensures that a good link is maintained between the cell station and the mobile phone virtually all the time. Obviously there will be some areas which are not easily covered, but these are normally comparatively few in most well-populated areas.

Control and signals

The cellular phone requires two types of channel to operate. Obviously the voice or signal channel is required to carry the signal information. In addition to this control channels are required. These are used to send information to and from the mobile phone to monitor its status while it is idle (switched on but not in use), to initiate a call and allocate the voice channel. As the phones are fully duplex they require one channel in each direction, and for each function. Thus a total of four channels is required, as shown in Figure 10.4.

In an analogue system the voice signal is a narrow-band FM signal. This gives good immunity to general noise and signal strength variations. The capture effect also gives good immunity to interference from distant cells. All of these aspects make it the ideal choice of modulation for this type of application.

More problems are encountered with the signalling side functions. Here multipath propagation, where the signal may reach the receiver via different

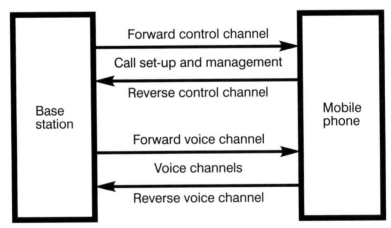

Figure 10.4 *Control and voice channels required for an analogue system*

paths and at slightly different times, means that errors are often present. Corruption becomes worse when signal strengths are low. To overcome this various forms of error detection and correction are employed.

Digital systems

The first mobile phone systems which were launched used analogue modulation techniques for the voice channel. This had a number of disadvantages. Speech quality was not always very good, and in addition to this it is very easy for people using scanners to listen to the conversations. Although this is illegal it does not stop those wanting to listen in from doing so.

There are two main types of digital network. The first is the GSM (Global System for Mobile Communication) system which operates at around 900 MHz. The other is the personal communications system (PCN) and this operates at the higher frequencies of around 1800 MHz

The idea behind GSM is that it uses an internationally agreed standard, and it is possible to use a phone from an operator in another country, for example if the user goes abroad. However, for this to occur, there must be a licence agreement between the operators of both countries. At the moment this is beginning to occur and it is possible to use GSM telephones in a variety of countries, and the number is continually increasing.

Table 10.1   Summary of cellular systems available

AMPS	Advanced Mobile Phone System (USA)
ADC	American Digital Cellular (works within the AMPS frequency plan)
C-NET	An analogue network (Germany)
E-TACS	Extended TACS (TACS-based system offering additional channels to accommodate higher number of users)
DCS	Digital communication system
GSM	Global System for Mobile communications. Digital system specified by CEPT – Committee of European Posts and Telecommunications
JDC	Japanese Digital Cellular
JTACS	Japanese Total Access Communications System – similar to TACS
NMT	Nordic Mobile Telephone – number following abbreviation refers to the band in use
TACS	Total Access Communication System (UK)

Table 10.2   Cellular phone system specifications

System	Rx Freq. MHz	Tx Freq. MHz	Channel Spacing kHz	Modulation
AMPS	869–894	824–849	30	FM
E-TACS	916–949	871–904	25	FM
NMT-450	463–468	453–458	25	FM
NMT-900	935–960	890–915	12.5	FM
GSM	925–960	880–915	200 (8)	GMSK
DCS-1800	1805–1880	1710–1785	200 (8)	GMSK

Figures in brackets refer to the number of users per channel

Private mobile radio (PMR) systems

Although the use of cellular phones has expanded very rapidly since the first systems were introduced, many private mobile radio systems (PMR) are still used. These systems are often used by businesses for keeping in contact with units out on the road. These systems are often used by taxi firms or service companies where frequent contact is required.

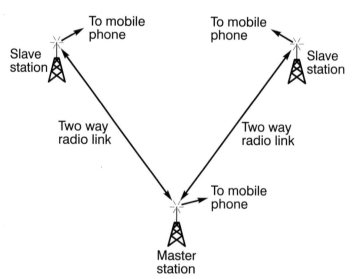

Figure 10.5 *A typical PMR radio system with a master and outlying base stations*

Systems in the UK are based on a 12.5 kHz channel spacing and can be found in sections of the VHF and UHF spectrum. FM is generally used along with double frequency simplex operation, i.e. only one person may speak at any given time, and the base station and mobile units use different frequencies.

As PMR systems are far more localized than cellular systems coverage is more limited. However, to improve the coverage over that which might be obtained with just one station, a system may consist of a master station with one or more slave stations in outlying areas. By linking them all together the coverage may be considerably increased, especially in hilly areas where a single station would have a very limited coverage area.

In early PMR systems the mobile station would be required to have the receiver on all the time, and it would be necessary to listen to all the messages to determine if one was required for any given station. Now it is possible to selectively call a particular station. This is done by sending a combination of five tones. The receiver is programmed to recognize these tones and only open the squelch when activated by these tones. In this way the receiver may appear dormant when messages not directed to it are not present.

Appendix:
Basic calculations

A large number of radio specifications involve the use of decibels. It is simply a ratio between two power levels using logarithms to the base ten. Although the basic unit is the bel, this is too large for most purposes and the decibel or dB, which is a tenth of a bel, is normally adopted.

Natural or Napierian logs can be used in the same way. In this case the units are nepers, but these are seldom used.

The number of decibels N can be calculated as below:

$$N = 10 \log_{10} \frac{P_2}{P_1}$$

where P_1 and P_2 are the power levels being compared.

If these levels are expressed in terms of voltages V_1 and V_2 then the formula changes

$$N = 20 \log_{10} \frac{V_2}{V_1}$$

Similarly for two currents

$$N = 20 \log_{10} \frac{I_2}{I_1}$$

If the impedances of the two points compared are different then this also has to be taken into consideration because a larger voltage is needed to dissipate the same power if the resistance is higher.

$$N = 20 \log_{10} \frac{V_2}{V_1} + 10 \log_{10} \frac{Z_1}{Z_2}$$

Z_1 and Z_2 are the impedances at the points where V_1 and V_2 were measured.

Table of decibels against power and voltage or current ratios

dB	Power Ratio	Voltage or Current Ratio
0.1	1.023	1.012
0.2	1.047	1.023
0.3	1.072	1.035
0.4	1.096	1.047
0.5	1.122	1.059
0.6	1.148	1.072
0.7	1.175	1.084
0.8	1.202	1.096
0.9	1.230	1.109
1.0	1.259	1.122
2.0	1.585	1.259
3.0	1.995	1.413
4.0	2.512	1.585
5.0	3.162	1.778
6.0	3.981	1.995
7.0	5.012	2.239
8.0	6.310	2.512
9.0	7.943	2.818
10.0	10.000	3.162
20	10^2	10.000
30	10^3	31.623
40	10^4	100.00
50	10^5	316.23
60	10^6	1000.0
70	10^7	3162.3
80	10^8	10 000
90	10^9	31 623
100	10^{10}	100 000

Index